A TEXTBOOK OF PLANT BIOLOGY

British Library Cataloguing-in-Publication Data
A catalogue record for this book is available from the
British Library

PLATE I

s

BEECHWOOD ON THE NORTH DOWNS

The ground vegetation consists chiefly of Dog's Mercury (*Mercurialis perennis*). The beech tree in the foreground has been attacked by a parasitic fungus. (*Polyporus adnetus*) which damages the wood and causes the branches to break off as shown in the photograph. Hence the common name, "Snapbeech" disease. The small bracket-like fructifications (*sporophores*) of the fungus are seen at *s*.

A TEXTBOOK
OF
PLANT BIOLOGY

BY

W. NEILSON JONES, M.A., F.L.S.
PROFESSOR OF BOTANY IN THE UNIVERSITY OF LONDON
(BEDFORD COLLEGE)

AND

M. C. RAYNER, D.Sc.
LATE LECTURER IN CHARGE, DEPARTMENT OF BOTANY, UNIVERSITY
COLLEGE, READING

Botany

The term 'botany' comes from the Ancient Greek word
botanē, meaning 'pasture', 'grass', or 'fodder', in turn derived
from *boskein*, meaning 'to feed or graze'. It chiefly involves
the study of plant life, as a branch of biology.

Traditionally, botany has also included the study of
fungi and algae by mycologists and phycologists respectively,
with the study of these three groups of organisms remaining
within the sphere of interest of the International Botanical
Congress. Nowadays, botanists study approximately
400,000 species of living organisms of which some 260,000
species are vascular plants and about 248,000 are flowering
plants.

Botany originated in prehistory as herbalism with the
efforts of early humans to identify – and later cultivate –
edible, medicinal and poisonous plants, making it one of the
oldest branches of science. Examples of early botanical works
have been found in ancient texts from India dating back to
before 1100 BCE, in archaic Avestan writings (an
Iranian language known only from its use in Zoroastrian
scriptures), and in works from China before it was unified in
221 BCE.

Modern botany traces its roots back to Ancient Greece,
specifically to Theophrastus (c. 371–287 BCE), a student
of Aristotle who invented and described many of its
principles. Today, he is widely regarded in the scientific

community as the 'Father of Botany'. Theophrastus's major works, *Enquiry into Plants* and *On the Causes of Plants* (both looking at plant structure, variety, reproduction and growth), constitute the most important contributions to botanical science until the Middle Ages, almost seventeen centuries later. Another work from Ancient Greece that made an early impact on botany is *De Materia Medica*; a five-volume encyclopaedia about herbal medicine written in the middle of the first century by Greek physician and pharmacologist Pedanius Dioscorides. *De Materia Medica* was widely read for more than 1,500 years subsequently.

Medieval physic gardens, often attached to monasteries, contained plants of great medicinal importance. They were forerunners of the first botanical gardens attached to universities, founded from the 1540s onwards. In the mid-sixteenth century, botanical gardens were founded in a number of Italian universities – and the Padua botanical garden in 1545 is the first of such, still in its original location. These gardens continued the practical value of earlier 'physic gardens' of the monasteries, and further supported the growth of botany as an academic subject.

Botanic gardens encouraged the work of academics such as the German physician Leonhart Fuchs (1501–1566). Fuchs was one of 'the three German fathers of botany', along with theologian Otto Brunfels (1489–1534) and physician Hieronymous Bock (1498–1554). Fuchs and Brunfels broke away from the tradition of copying earlier works to make original observations of their own, whilst Bock created his own system of plant classification. In 1665, using an early microscope, another famed botanist, the Polymath Robert

Hooke (1635 - 1703) discovered 'cells' in plant tissue, a term he coined. During this early period, lectures were also given about the plants grown in the specially constructed botanic gardens, and their medical uses demonstrated. Botanical gardens came much later to northern Europe – largely due to the obvious differences in temperature. The first in England was the University of Oxford Botanic Garden, constructed in 1621.

Efforts to catalogue and describe the collections of these gardens were the beginnings of plant taxonomy, and led in 1753 to the binomial system of Carl Linnaeus that remains in use to this day. Linnaeus's system was a hierarchical classification of plant species providing a solid reference point for modern botanical nomenclature. This established a standardised binomial or two-part naming scheme where the first name represented the genus and the second identified the species within the genus. For the purposes of identification, Linnaeus's *Systema Sexuale* classified plants into twenty-four groups according to the number of their male sexual organs. The twenty-fourth group, *Cryptogamia*, included all plants with concealed reproductive parts, mosses, liverworts, ferns, algae and fungi.

Increasing knowledge of plant anatomy, morphology and life cycles led to the realisation that there were more natural affinities between plants than Linnaeus had indicated however. Scholars such as Adanson (1763), De Jussieu (1789), and Candolle (1819), all proposed various alternative natural systems of classification that grouped plants using a wider range of shared characters and were extensively followed. The Candollean system reflected his ideas of the progression of morphological complexity and were developed the classifications by Bentham and Hooker, influential until the mid-nineteenth century. Darwin's publication of the *Origin of Species* in 1859 and his concept of common descent, further required modifications to the Candollean system to reflect evolutionary relationships as distinct from mere morphological similarity.

In the nineteenth and twentieth centuries, new techniques were developed for the study of plants, including methods of optical microscopy and live cell imaging, electron microscopy, analysis of chromosome number, plant chemistry and the structure and function of enzymes and other proteins. In the last two decades of the twentieth century, botanists exploited the techniques of molecular genetic analysis, including genomics and proteomics and DNA sequences to classify plants more accurately. Particularly since the mid-1960s there have been advances in understanding of the physics of plant physiological processes such as transpiration (the transport of water within plant tissues), the temperature dependence of rates of water evaporation from the leaf surface and the molecular diffusion of water vapour and carbon dioxide through stomatal apertures.

Twentieth century developments in plant biochemistry have been driven by modern techniques of organic chemical analysis, such as spectroscopy, chromatopgraphy and electrophoresis. With the rise of the related molecular-scale biological approaches, the relationship between the plant genome and most aspects of the biochemistry, physiology, morphology and behaviour of plants can be subjected to detailed experimental analysis. Such developments have enabled advances in areas as diverse as pesticides, antibiotics and pharmaceuticals, as well as the practical application of genetically modified crops designed for traits such as improved yield.

The study of botany is an incredibly important science, as plants underpin almost all life on earth. They generate a large proportion of the oxygen and food that provide humans and other organisms with aerobic respiration with the chemical energy they need to exist. In addition, they are influential in the global carbon and water cycles and plant roots bind and stabilise soils, preventing soil erosion. Plants and the science of botany are crucial to the future of human society – allowing insight into our food, natural environment, medicine and products. It is a branch of human endeavour with an incredibly long and varied history, and it is hoped the current reader enjoys this book on the subject.

PREFACE

THIS is not a general textbook of botany; a good excuse would be required for adding yet another to the number of excellent books of this class already on the market. The authors hope, however, that the present volume will meet a need which they believe is not entirely satisfied.

The ordinary elementary textbook of botany is intended to serve as an introduction to the study of plants and provide a basis for more advanced work. In practice this more extended study is in many cases not pursued by the student, with the result that he leaves the subject before he has realised its wider biological aspects. This is regrettable, inasmuch as school botany is for many the only opportunity presented for acquiring a grasp of biological principles, and it would appear desirable that the subject should be taught in such a way as to provide the future citizen with a general grounding in biology, without assuming that the elementary student of science must needs develop into the advanced student or the professional scientist. The aims which the authors have endeavoured to keep before them are: firstly, to design an elementary course which will serve as an introduction to the scientific method; secondly, to enable the student or intelligent layman to acquire an understanding of the relation of plant life to general biological knowledge; and thirdly, while fulfilling these objects, to provide a botanical textbook for the use of the senior classes of schools and junior classes of the University.

Many of the modern developments of botany are regarded as "advanced" subjects, but the authors have sought to present the essential significance of certain of these in simple language so as to indicate their position in the general scheme of life, and to do so in a form attractive and intelligible to the elementary student. It is not claimed, be it understood, that this volume is "light" reading, but pains have been taken to present the matter as clearly as is permitted by the nature of the subject and the limitations of space.

It will be found that certain aspects have been sacrificed in order to include others of wider interest. The attention of the student is focused upon the living plant—upon the

manner in which it carries on its life processes, upon its reproduction and evolution and upon its relations with its surroundings.   As a result, no complete treatment of anatomy
or morphology has been attempted, but reference made only
to such features as are essential to a clear understanding of
vital processes.   The authors have felt the less concerned at
these omissions in view of the fact that these parts of the
subject are treated adequately in many admirable textbooks,
and they have felt justified in assuming an elementary
knowledge of morphology as part of the equipment of the
student.   They are fully conscious that many important
problems have been ignored or treated all too briefly.   It
would have been desirable, for example, to have included a
section dealing with the correlation between structural modification and environment, and another dealing with the
economic aspects of plants, and yet others.   The inclusion
of such additional matter, however, would have resulted in a
volume considerably larger than the present, which already
exceeds the limits set.

It is hoped that the practical exercises will prove useful
to the teacher as well as to the student.   They are intended
to be suggestive rather than comprehensive.

The need for curtailing size and expense of production had
also to be borne in mind in regard to illustrations.   The plan
followed has been to exclude illustrations of material readily
accessible to the student, except in so far as diagrams have
served to make clearer the meaning of the text.

For permission to use Plate VI. we are indebted to the
courtesy of Dr. E. P. Farrow and the editor and publishers
of the *Journal of Ecology;* and for the diagram on p. 117 to that
of Dr. F. J. Lewis and the Royal Society of Edinburgh.

The other plates are from original photographs; the text
figures and diagrams, with a few minor exceptions, have been
drawn specially for the present volume.

We also take this opportunity of acknowledging the kindness
of botanical and literary friends who have read and criticised
the proofs.

W. N. J.
M. C. R.

BEDFORD COLLEGE, LONDON
1920

CONTENTS

LIST OF PLATES

A TEXTBOOK OF PLANT BIOLOGY

INTRODUCTORY*

THE PLANT AS MACHINE. REPRODUCTION. THE PLANT IN
RELATION TO ITS SURROUNDINGS

An organism—whether plant or animal—may be likened in
many ways to a machine: it has the same limitations as a
machine in so far as it requires to be supplied with energy
in some form in order to function, and then will only do so
under specific conditions. Living organisms differ from
machines, however, in the powers they possess of adapting
themselves to changes in the environment and in their capacity
of self-reproduction.

Botany or the study of plants possesses, therefore, at
least three aspects: The plant as a machine; reproduction,
in its widest sense; the correlations between the plant and its
physical and biological environment.

THE PLANT AS A MACHINE

Machines differ from one another in the forms of energy
with which they work and in the vehicles they employ for
transmission of the energy. Thus, a windmill is a mechanical
engine using the air as the vehicle of its energy, a steam engine
is a form of heat engine using water for conveying its energy,
while a plant makes use, for the most part, of chemical energy
of which the expenditure is controlled by the protoplasm.

Now, a steam engine relies on some immediate external
source of energy, usually coal, wood, or oil, from which on

---

* The student is advised to read this chapter again when the subsequent chapters have been studied.

I

combustion the energy is liberated in the form of heat. In this respect " plant engines " are of two kinds, those which derive their energy directly from sunlight, and those which do not. The former class of plants must obviously be equipped with some mechanism whereby solar energy may be absorbed. To this end they are provided with light-absorbing or *assimilating* pigments, much the most important of which is the green pigment *chlorophyll,* although other assimilatory pigments occur, as for instance in the Brown Seaweeds and Red Seaweeds.

The energy cycle of a green plant is briefly of the following nature. A certain portion of the light energy from the sun is absorbed by the chlorophyll and is then utilised in building up organic compounds, of which carbohydrates such as sugar and starch form the bulk. When these carbohydrates are oxidised (or otherwise decomposed), they again give out energy which is made use of in carrying out the various life processes of the plant. The liberation of this stored or *potential* energy is controlled by the protoplasm. The green plant is thus driven by the sun and is kept running by virtue of its capacity to absorb and store up solar energy. Plants which do not contain chlorophyll, or a similar pigment, are unable to absorb solar energy and are compelled to use the energy stored up in chemical compounds manufactured by plants containing these pigments, upon which they are, therefore, ultimately dependent. The former constitute the groups of plants known as *saprophytes* or *parasites* according as to whether they make use of the organic compounds contained in dead or in living organisms. Green plants, obtaining their energy directly from the sun, are contributors to the world's supply of available energy, in that they produce a larger amount of organic compounds than they need for their individual requirements: saprophytes and parasites reduce the world's supply by living at the expense of the chemical compounds produced by green plants. The whole animal kingdom, it should be observed, live saprophytically or parasitically as regards their energy requirements. The capacity on the part of green plants for absorbing solar energy and storing it up in the form of organic compounds is one of which the importance cannot be over-emphasised, since the possibility of life as we know it depends absolutely upon it. Indeed, were it not for the existence of green plants or other organisms capable of utilising solar energy,—perhaps one might say of the green colouring-matter chlorophyll,—life would be impossible. The fundamental importance of agriculture turns upon the fact that the object of agricultural practice is to add to the world's available supply of energy by the efficient use of green plants for the collection of solar energy, and the storage of this energy, either in plants or animals, for the use of human beings.

But one may go even further than this. With the exception of the comparatively small amount of energy derived from watermills and windmills, the activities of green plants are responsible for the whole of the energy used for industrial purposes, which is derived for the most part from that stored up in coal, oil, wood and peat. The importance, even from a practical standpoint, of a subject which is concerned with the processes which have rendered possible the building up of our civilisation, and indeed our very existence, need not be laboured.

In dealing with the plant machine two complementary aspects must be kept in view: the construction of the machine and the manner of its working, i.e. the *anatomy* and the *physiology* of the organism. Since an understanding of the one is necessary for the full appreciation of the other, we shall endeavour to treat the two aspects concurrently so far as this is possible.

REPRODUCTION

The capacity for *reproduction* possessed by even the most complicated piece of machinery is of an entirely different kind from that possessed by living organisms. An automatic machine may produce in large numbers and almost untended, say, a particular kind of screw-nail, but it is unable to produce another automatic machine like itself: a printing press may *produce* a large number of copies of some book, but cannot *reproduce* itself. On the other hand, a living organism in the process of reproduction gives rise to other organisms like itself. A particular kind of cell tends to produce another cell like itself, while on the larger scale a complex multicellular Flowering Plant tends to produce another plant which is (or will become) like that from which it arose.

There is also a contrary tendency of *variation* to be recognised in the reproduction of living things. In the individual organism this finds expression in the dissimilarities between parents and offspring; although the offspring resemble the parents they are not, as a rule, exactly alike in all particulars. Then, again, in the cell divisions that occur during the growth of a single multicellular individual towards maturity, cells which are all alike originally give rise to others unlike themselves, thus producing eventually that differentiation of the tissues which renders the structure of the mature organism so complex.

The term " reproduction " thus includes phenomena covering a wide field. The most fundamental of these is the process of *cell-division*, whereby one cell produces an exact counterpart of itself. These cell-divisions involve the division of the structures within the cell and, in the case of the nucleus, this is attained by means of an extremely elaborate mechanism, which ensures that the two portions into which it is divided shall be qualitatively equal. On the other hand, each cell of the growing region of a multicellular plant gives rise to others which may differ among themselves. A study of reproduction includes, therefore, the manner in which plant tissues originate, the way in which new tissues or organs become differentiated, and so on. But besides these questions, relating to divisions of cells considered as units, there is the broader question of the reproduction of the multicellular individual as a whole. In this connection it must be realised that while each cell of a multicellular individual behaves and may be treated in many ways as an independent unit, there is, nevertheless, a relation or correlation between them all that gives to the whole complex of cells a joint individuality. The relations between the cell and the individual of which it forms a part may be compared, for purposes of illustration, with the unit individuality of the shareholders, directors, etc. of a company and the composite individuality of the company they compose.

The types of reproduction that occur in the individual plant fall into two categories, sexual and asexual. Under the latter are included such natural means of vegetative increase as is seen in bulbs, rhizomes, etc., and also such as can be induced artificially by the horticultural operations of grafting, making cuttings and so forth. A study of sexual reproduction is concerned with the process of fertilisation and its results; with the way in which the sex-cells or gametes are formed, their union at fertilisation, and the manner in which the fusion-product or *zygote* develops into the embryo. In highly specialised plants, such as the Flowering Plants, there are, furthermore, many facts and problems centring round the flower and pollination, to the study of which an immense amount of labour has been devoted, perhaps more than is justified when compared with that spent on more fundamental problems, although the interest and fascination of the subject is not to be denied.

Lastly there is the subject of heredity, which has made such notable advances in recent years and which promises to become of great practical value in many directions in the future, not only in the elucidation of problems of inheritance, but also in providing information as to the fundamental attributes of living matter, and in throwing light on the nature of organic evolution.

THE PLANT AND ITS ENVIRONMENT

In this brief review the plant has been imagined as though it were an isolated organism, living under uniform conditions, whereas, in point of fact, it is surrounded by a variety of constantly changing physical and biological influences. The remarkable capacity of adapting themselves to new surroundings possessed by living organisms differentiates them from machines, just as does the character of their reproductive processes.

We find plants, in common with animals, can react to changes in their surroundings; in scientific language, they are sensitive to certain external stimuli to which they react in various ways. Among the more obvious of such reactions are the movements that occur in response to the stimuli of gravity, light, touch, etc. Among the lower plants there may be movement of the whole organism; in the higher plants, the movement is localised in certain organs and may take place either with considerable rapidity, or as the result of slow growth. The movements of the algal plant *Chlamydomonas* in response to variations in light intensity; the rapid contraction of the stamens of many Composite flowers, when the sensory region at the base is touched; the gradual growth of a horizontally placed stem into a vertical position under the influence of gravity, serve as illustrations of this type of reaction.

In addition to reactions rendered more or less obvious by movement, the whole manner of growth and life of the plant may be changed in greater or less degree when the physical environment is altered, and the changes that occur are almost invariably such as to render the plant better adapted to the conditions under which it is growing. A large amount of experimental work has been carried out to determine the extent to which plants are able to respond to alterations in the surrounding conditions and the effect of such alterations upon the life processes.

Alterations of the environmental conditions may lead, not only to changes in the individual plant, but also, in time, to the appearance of plant types showing special characteristics. Exactly how this race modification is brought about need not be discussed now; the only point with which we are immediately concerned is that besides adaptation of the individual to changed conditions, there is also adaptation of the race. Thus, it is found that plant species inhabiting situations abnormally dry, damp, windy, etc., often show adaptations to the special conditions of the locality, and it becomes convenient to divide plants into groups such as *xerophytes*, *epiphytes*, *climbers*, etc.

What has been said with reference to the physical conditions of the environment applies equally to the biological influences. The reaction of one organism upon another has received increased attention of late, and the immense importance of this field of investigation is now recognised.

The effect of one organism upon another may be exercised through chemical or physical agencies, or may be due to a direct protoplasmic reaction. Thus, a tree affects the vegetation beneath it by cutting off a considerable portion of the light that would otherwise be available; the presence of a particular species of insect may be of importance to a species of plant by pollinating its flowers; the bacteria of the soil affect the vegetation growing thereon by altering soil fertility, and so on. The protoplasmic reactions of one organism upon another, resulting from the mutual reactions of their respective protoplasms, are much more subtle and less easily understood.

Upon such reactions depend what is called the *immunity* or *susceptibility* of an individual or a species to attack by parasites; the possibility of grafting one species upon another or of crossing one species with another; the occurrence of so-called *symbiotic partnerships* between two organisms; and many other phenomena which at first sight seem to have little in common, but are linked together by the fact that they are the expression of mutual reactions on the part of living organisms. The study of plants in their natural habitats, including the effects of external conditions generally and their interaction with one another, is known as *Plant Ecology*, and is one of the most exacting aspects of Botany, in that it necessitates an extensive and thorough knowledge of other branches of Botany, together with a considerable acquaintance with other Natural Sciences.

In this introductory chapter we have endeavoured to outline the field of inquiry that lies before the student. It can hardly be complained that this is limited in extent; more likely is the student to find its range somewhat bewildering at first. It is hoped, however, that the arrangement adopted will enable him to link together the different aspects of the subject into an harmonious whole, and that it will provide an introduction to the more important fundamental conceptions of plant life and their relations with human interests.

PART I—THE PLANT AS MACHINE

CHAPTER I

THE CELL

PROTOPLASM ; CHEMICAL AND PHYSICAL PROPERTIES. THE PROPER-
TIES OF LIVING MATTER ; MECHANISTIC AND VITALISTIC THEORIES.
THE CELL ; NUCLEUS, CYTOPLASM, PLASTIDS, ETC. TISSUES. THE
PHYSICAL PROPERTIES OF THE LIVING CELL; PERMEABLE AND SEMI-
PERMEABLE MEMBRANES ; OSMOSIS ; OSMOTIC PRESSURE ; TURGOR ;
PLASMOLYSIS. THE NATURE OF THE PROTOPLASMIC MEMBRANE

Protoplasm—All manifestations of life, whether in the animal
or vegetable kingdom, are bound up with the presence of the
living substance *protoplasm*. Before we begin our study of
the plant machine it will be well, therefore, to inquire briefly
as to the nature of this all-important substance.

As a result of chemical analysis, it is found that protoplasm
consists largely of *proteins*, a class of complex organic substances containing the elements *carbon, hydrogen, oxygen,
nitrogen*, often combined with *sulphur* and *phosphorus*, and
giving characteristic chemical reactions. The presence of a
number of other chemical substances will also be revealed
by analysis; but since these bodies occur in the food of the
protoplasm and as products of its activities, it is impossible
to say from the analysis to what extent they form an essential
part of protoplasm itself. Moreover, active protoplasm contains a large percentage of water, although, in the dormant
state, as in seeds, the amount of water may be quite small.

We must be cautious, however, in drawing conclusions as to
the constitution of protoplasm from the results of chemical
analyses. All that we can be certain of is that the dry substance of *dead* protoplasm consists chiefly of protein substances :
the nature of *living* protoplasm eludes direct analytical attack

9

by chemical methods, since the very first step in the analysis inevitably causes death. The methods of the chemist enable us to discover the chemical elements present in protoplasm and to recognise to some extent the combinations in which they occur; but they give us no direct information as to the manner in which these materials are combined together in living protoplasm.

The proteins themselves are bodies of very complex chemical structure. They can be disintegrated by chemical methods into a series of substances of decreasing complexity, many of which are actually found in the plant and represent stages in the building-up or breaking-down of proteins in the " plant laboratory." It has long been the dream of the chemist to reverse this procedure and to build up these protein substances in his laboratory from inorganic materials. Certain of the simpler substances formed when proteins are broken down chemically have been so constructed, and there seems no theoretical reason why the chemist should not in time succeed in artificially building up the more complex organic substances we call proteins; but even were this possible, he would still be far from reversing the first step of the analysis and compounding these artificially-made proteins to form living protoplasm.

We may note in passing that two views are held by scientists regarding the nature of life. According to one view it is believed that the unique properties of protoplasm depend on the character of the materials composing it and the way in which they are combined together. This may be stated in another way by saying that if the chemist could manufacture all the constituents of protoplasm and could bring them together under appropriate conditions, there is no reason why he should not produce in his laboratory a substance which would manifest the phenomena of life. This is called the *mechanistic* view, and those who hold it believe that the difference between living and non-living material is one of degree rather than one of kind. Those who hold the contrary or *vitalistic* theory of the nature of life believe that the difference between living and non-living material is a fundamental one, and that living organisms, however simple in nature, possess, over and above any peculiarities of chemical or physical constitution, a mysterious " vital principle " to the possession of which the manifestations of life are due. Speculations as to the nature of this vital force belong to the realm of philosophy rather than to that of natural science.

Let us turn, now, to a consideration of what may be learnt by direct observation of the nature of living protoplasm. Comparatively large masses of free protoplasm are found in certain stages of the life history of the " Slime Fungi " or *Myxomycetes*, a group of lowly organisms which show features common to both animals and plants. In the active or vegetative stage a Myxomycete consists of a mass of protoplasm which, under certain conditions, emerges from the wood or other material in which it is growing. This *plasmodium*, as it is called, provides us with material for studying some of the more fundamental characteristics of protoplasm. We find, for instance, that it reacts to changes in the external conditions and that flowing or creeping movements can be induced and directed by moisture, light, and other external influences.

To the naked eye a plasmodium is a mass of slimy, rather opaque substance,—white, creamy, orange, or otherwise coloured according to the species. By appropriate treatment it is possible to cause an active plasmodium to spread itself out on a glass slide so as to form a delicate network of protoplasm which can be examined microscopically. It can then be noted that the opacity and colour is due to the inclusion of solid particles taken in or manufactured by the plasmodium, but by examination of the clear borders of the network we may gain an insight into the structure of the protoplasm free from such inclusions. The protoplasm in this region of the plasmodium appears to the naked eye to be uniform and semitransparent. Examined by high powers of the microscope, it is resolved into a clear substance, the *cytoplasm*, in which numerous minute granules or droplets are suspended; scattered throughout the mass are small denser portions or *nuclei*.*

The Cell—Now observation shows that if the living protoplasm is divided into separate portions, each can continue its

---

* It may be noted that the plasmodium of a Myxomycete, shortly after emergence into the light, passes from the active, vegetative phase into the reproductive phase, in which condition it appears as characteristic spore-cases or sporangia, the characters of which differ with the species. The latter is the condition in which these organisms are most commonly found.

independent existence, provided it contains at least one nucleus. In point of fact, the Myxomycete, at one stage of its existence, occurs naturally in the form of separate fragments of protoplasm, each of which consists of a single nucleus with its surrounding cytoplasm; large numbers of these separate portions subsequently coalesce to form the plasmodium.

Thus we reach the conception that the living unit is a nucleus with its accompanying cytoplasm, each such unit being known as a *cell*. The mass of protoplasm forming the

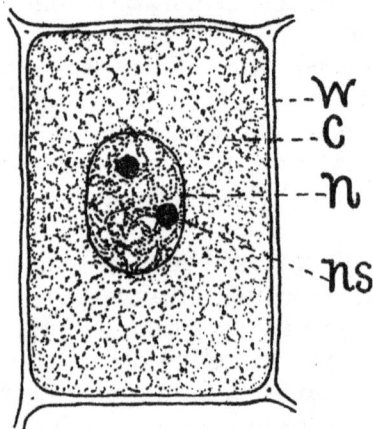

FIG. 1—CELL FROM GROWING REGION OF BEAN ROOT TO ILLUSTRATE MICROSCOPIC STRUCTURE OF YOUNG PLANT CELL. (Highly magnified)
w, cell-wall; *c*, cytoplasm; *n*, nucleus; *ns*, nucleolus

plasmodium may therefore be regarded as an aggregate of numerous cells, which, although they form the plasmodium as a whole, are yet to be considered as independent units to the extent that any one of them is capable of carrying on an independent existence.

In the plasmodium of a Myxomycete, the cytoplasm around one nucleus merges indistinguishably into the cytoplasm of those surrounding it, and it is impossible to say where one cell ends and the next begins. This condition is exceptional in plants; each plant cell usually excretes around itself a delimiting wall of *cellulose* which marks off definitely one cell from the next. Now, since each cell or portion of living protoplasm forms its own enveloping cell-wall, the wall separating

any two adjoining cells belongs partly to one cell and partly
to the other. The two portions making up the dividing
wall can be made evident by appropriate staining or other
means, since the middle region, where the portions contributed by each cell meet, gives different chemical reactions
and has different staining properties from the rest of the
wall.

In the light of the above let us turn to a consideration
of the minuter structure or *histology*, as it is called, of a multicellular plant such as a Flowering Plant. In the roots and

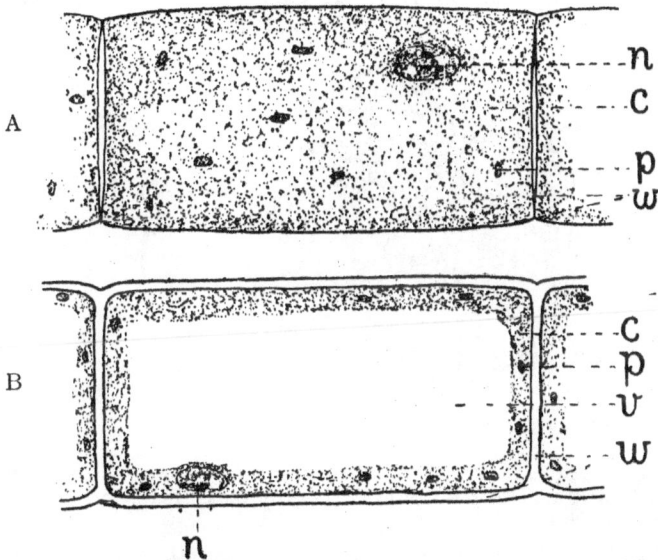

FIG. 2—CELLS FROM HAIR OF FOLIAGE LEAF OF *Primula sinensis* TO
ILLUSTRATE MICROSCOPIC STRUCTURE OF A MATURE PLANT CELL.
(Highly magnified.) A, surface view; B, optical section

w, cell-wall; *c*, cytoplasm; *n*, nucleus; *p*, plastid; *v*, cell-vacuole with
cell-sap

stems of Flowering Plants the structure typical of young
cells can best be observed at the tips or growing points where
new cells are continually being formed. These growing
points are composed of cells separated from each other by
delicate though definite cellulose walls. The minutely granular
protoplasm, in which is embedded the nucleus, completely
fills each cavity. The nucleus itself has an elaborate structure
into the details of which we need not now enter. As these

young cells become older they grow rapidly in size. Their growth in size is, in fact, more rapid than the growth in quantity of the protoplasm, so that in a short while the amount of the latter is insufficient to fill completely the cavity within the enlarged cell-wall, and spaces appear filled with liquid or *cell-sap*. As the volume enclosed by the cell-wall increases still further, these spaces or *vacuoles* increase in number and size, and may eventually coalesce to form a large cavity,— the *cell vacuole*,—occupying the middle of the cell. This condition, in which the protoplasm completely surrounds the cell-sap like a sack or bag, and is enclosed externally by the

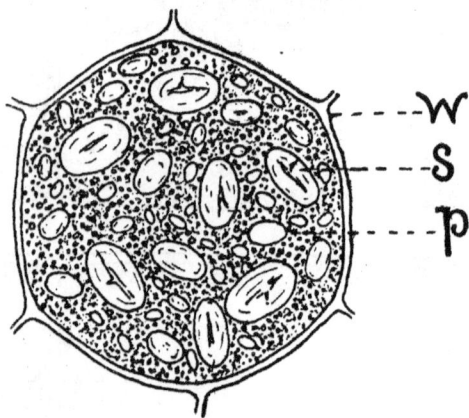

FIG. 3—CELL CONTAINING RESERVES FROM COTYLEDON OF EMBRYO BEAN PLANT IN RESTING SEED. (Highly magnified, and somewhat diagrammatic)

w, cell-wall; *s*, starch grain; *p*, protein reserve (aleurone grains)

cell-wall, is typical of *living* plant cells in the mature condition. In old cells, owing to increased size, there may be only sufficient protoplasm to form a very thin layer around the central vacuole, and it may be difficult to distinguish it from the cell-wall, to which it is closely applied, unless special means are adopted to render it conspicuous. In mature cells, too, the cell-wall frequently becomes thickened, and may contain other substances in addition to cellulose (figs. 1, 2).

Included within the cell protoplasm are usually bodies which may be subdivided into two kinds: specialised portions of the protoplasm, of which the most conspicuous and important are the *plastids*; non-living substances formed by the protoplasm, some of which are insoluble and are present either as solid granules, e.g. starch, or as droplets, e.g. oil, whilst others are soluble and are present in solution in the protoplasm and cell-sap (fig. 3).

We must think of the cell as a kind of workshop or laboratory in which the activity of the protoplasm is responsible for the constant manufacture and alteration of a vast number of substances. Some of these go to form new protoplasm; some are excreted and serve to build up the permanent framework of the plant, as does the cellulose of the cell-walls; others, such as starch, play an important part in the processes of nutrition and growth; a few are of the nature of waste products which, since plants have no method of getting rid of them, accumulate in the cells.

In different regions of the plant the living cells are of different shapes, the cell-walls differ in thickness, form, and chemical composition; the auxiliary bodies such as plastids and starch grains may differ also from one another; but all *living* cells have essentially the structure described above in that they possess protoplasm, consisting of cytoplasm and nucleus, a central vacuole (except in very young cells), and a cell-wall the basis of which is cellulose. A not inconsiderable part of the body of a Flowering Plant consists, moreover, of *dead* cells, the protoplasm of which has died and disappeared, leaving only the surrounding cell walls and possibly certain of the non-living contents. The term " cell " is still applied, as a matter of convenience, to these shells bereft of the living protoplasm.* Their usefulness to the plant is by no means at an end; they serve the purpose of strengthening and stiffening the plant body, of providing conducting channels for the rapid conveyance of water, and so forth.

The particular form assumed by the cells in any region of the plant is correlated with the special functions they perform. Cells carrying out like functions and consequently possessing similar structure are commonly associated together to form *tissues*, the whole plant body being thus built up of tissues of different kinds,—mechanical or skeletal tissues, water-conducting tissues, and so on.

* And as an historical survival. When plant tissues were first subjected to microscopical examination, it was the cell-walls that attracted most attention and suggested the term " cell." The proto. plasm was overlooked or regarded as of minor importance.

All plants do not have such diverse and complex tissues as the Flowering Plants; the cells of plants low in the scale of evolution may show but slight diversity of type among themselves. Indeed, in the simplest plants and animals, the whole organism consists of a single cell. In such unicellular plants all the life functions are performed by one cell, so that the possession of a special form on the part of a cell which carries out a particular function, as is characteristic of the higher plants and animals, must not be regarded as a necessity but rather as a convenience. The *physiological division of labour* which results from apportioning the life functions among groups of cells, allows the cells of each group to assume the form specially suited to their particular tasks.

Much of what has been said above applies equally to the animal kingdom. Animal cells are distinguished from vegetable cells chiefly by the absence of a firm, cellulose wall and of a large central vacuole around which the protoplasm forms a thin layer. Although this distinction is a very general one, it is not absolute: the young cells near the growing points in Flowering Plants, as we have seen, contain no central vacuole; the cell-walls separating one cell from another are sometimes not formed in plants, which leads to the formation of *multinucleate cells*. That there should be no absolute distinction between the cells of animals and plants is not surprising, in view of the fact that the protoplasm of both groups has had, presumably, a common origin.

Notwithstanding the fact that the whole mass of protoplasm forming a Flowering Plant is chambered off into individual cells by rigid cellulose walls, connection between component cells is not entirely severed. Minute openings are left in the cell-walls through which run fine strands of protoplasm which serve to maintain organic connection between adjoining cells. It is owing to this fact that the joint individuality of the whole complex of cells is able to manifest itself, while at the same time the movement of substances from one cell to another is facilitated.

It should be noted that, although in a Flowering Plant the rigid framework formed by the cell-walls prevents the protoplasm as a whole from executing creeping movements like those to be observed in the plasmodium of a Myxomycete, yet the protoplasm within each compartment is not so constrained, and flowing or streaming movements of the cytoplasm within the cells can easily be observed when suitable material is examined with the microscope. Changes in position of the nucleus, plastids and other bodies contained within the cell can also be observed.

Physical Properties of the Cell—In addition to its structural features, certain physical peculiarities of the cell are of the greatest importance: these will be more readily understood if we turn from the plant cell for a moment to consider the physical principles involved. We may commence with a generalised statement to the effect that *any change taking place in the relations between the parts of a system which involves an exchange of energy is always such as tends, on the whole, to lessen the difference between them.* Thus, if hot water is poured into a cold glass, the water will become colder and the glass hotter until the two are at the same temperature; if two vessels containing water at different levels are connected by a pipe, so that water can flow from one to the other, the change occurring will be such as to lessen the difference in levels; and so on. This will be found to be a useful guiding principle in the interpretation of the phenomena about to be described.

Imagine a solution of sugar completely enclosed by a membraneous bag that has the property of allowing water to pass through it readily but which is entirely impervious to the dissolved sugar. Such a membrane is known as a *semi-permeable membrane* as regards the sugar solution. Suppose, also, that the membrane has the further property of infinite extensibility, i.e. it does not offer resistance to being stretched to any extent. If such a bag is surrounded by a solution of sugar, weaker than that which it encloses, then, in conformity with the principle stated above, there will be a tendency for the solutions within and without the bag to approximate in strength. Since the membrane is impervious to sugar, this tendency can manifest itself only by passage of water through it; water will pass from the weaker solution through the membrane to the stronger solution within, until the two solutions are of the same strength. This will necessitate, of course, an increase in the size of the bag to accommodate the additional water. Conversely, if the outer solution is more concentrated than that inside, water

2

will pass from the inside to the outside, and the size of the bag be reduced until the concentrations on the two sides of the membrane are similar. The amount of alteration in size of the bag will depend, evidently, upon the degree of difference in concentration of the solutions within and without; the greater the difference in the concentrations, the greater the change in size of the bag.

Now let us consider what will happen if the membrane is completely inextensible. If the solution inside is stronger than that outside, then, as before, there is a tendency for water to pass from the weaker to the stronger solution,— from the outside to the inside of the bag. Since the volume of the bag is assumed to be unalterable, the pressure within must increase as soon as any additional water enters. As more water enters, the pressure still further increases until it reaches a value such that it just counterbalances the tendency for more water to enter. The greater the initial difference in concentration of the solutions inside and outside, the greater will be the pressure set up within the bag.

The force responsible for the passage of water through the semi-permeable membrane is known as *osmotic force,* and the value of the pressure produced within the bag, when the membrane is inextensible and the liquid outside is pure water, is said to be the *osmotic pressure* of the solution in the bag at the time. For example, the osmotic pressure of a 10 per cent. solution of cane sugar is about 97½ lb. per square inch (6½ atmospheres); so that the pressure produced within a bag containing a 10 per cent. solution of sugar, when the surrounding liquid is pure water, will be 97½ lb. per square inch, *assuming that the material of which the bag is composed does not stretch and is impermeable to sugar.*

Every solution possesses an osmotic pressure the value of which depends upon the nature of the substance dissolved, the liquid in which it is dissolved, the concentration of the solution, etc. When a solution, completely enclosed within a membrane permeable to the solvent but impermeable to the substance dissolved, is suspended in the pure solvent, the osmotic pressure is manifest, on the solution side of the membrane, as an actual pressure. The maximum value of this is equal to the osmotic pressure of the solution, and represents the pressure necessary to counterbalance the tendency for more water to pass through the membrane.

The osmotic pressure of any solution may be estimated directly by measuring the pressure obtained in this way, but the experimental difficulties are considerable and indirect methods are usually employed. The value of the osmotic pressure of a solution may also be determined theoretically from the fact that the osmotic pressure of a substance in solution is the same as the pressure it would exert if it could exist as a gas occupying the same volume.* In fact, in the case of dilute solutions, the general laws concerning the behaviour of gases can be applied to substances in solution. Thus, just as doubling the amount of gas in a given volume doubles the pressure, so doubling the concentration of a solution increases the osmotic pressure twofold, and so on. In the case previously considered, where there is a concentrated solution within the bag and a weak solution outside, the value of the actual pressure produced will be equal to the difference between the osmotic pressures of the solutions inside and outside the bag.

Actual membranes, such as a piece of sausage skin or vegetable parchment, unlike the hypothetical membranes we have been considering, are neither inextensible nor indefinitely extensible, but stretch within limits and resist extension the more they are stretched. In actual fact, a bag made of membrane, if containing a solution stronger than that by which it is surrounded, will take in a certain amount of

---

* As explained in chemistry textbooks, the pressure exerted by the molecular weight of any gas occupying a volume of 22·4 litres at 0° C. is 1 atmosphere or about 15 lb. per square inch. Cane sugar, $C_{12}H_{22}O_{11}$, has a molecular weight of 342; therefore 342 gms. of cane sugar (if cane sugar could exist as a gas), in a space of 22·4 litres would exert a pressure of 15 lb. per square inch; or a solution containing 342 gms. in 22·4 litres possesses an osmotic pressure of 15 lb. per square inch. A 10 per cent. solution contains 10 gms. in 100 c.c. or 342 gms. in 3420 cc.＝3·42 litres. A 10 per cent. solution is therefore 22·4/3·42＝6·5 times as concentrated as a gram molecular solution, and therefore has an osmotic pressure 6·5 times as great, namely, 6·5 × 15＝97½lb. per square inch. In making this calculation, it must not be forgotten that *electrolytes* in dilute solution are almost completely dissociated. A gram molecular solution of common salt exerts an osmotic pressure of 30 lb. per square inch instead of 15 lb. per square inch in consequence, since each molecule splits up into an ion of sodium and an ion of chlorine. The increased pressure due to dissociation must be allowed for in each case.

water and become stretched thereby. The solution contained in the bag does not absorb so much water as to reduce the concentration to that of the solution outside,—the concentration within will be just so much greater than that without, that the tendency for more water to enter by osmosis balances the pressure tending to force water out of the bag.

Thus, if a bag composed of a natural membrane, containing initially a 10 per cent. solution of sugar be immersed in water, the final pressure produced inside will not be the osmotic pressure of a 10 per cent. solution, but that of a somewhat weaker solution, since some water has entered the bag as a result of stretching.

Substances possessing the character of semi-permeability when in the form of a membrane are of frequent occurrence, and amongst them must be included living protoplasm.

Let us turn now to plants and consider the application of these physical facts to plant cells. A plant cell, from this point of view, consists of an aqueous solution of various substances,—the cell-sap,—surrounded by a semi-permeable membrane of a complicated kind,—the protoplasm.* From what has been said above, it is evident that if such a cell is surrounded by pure water or by a solution of salts weaker than the cell-sap, it will tend to increase in size owing to the entry of water. Since the protoplasm surrounding the vacuole is impermeable to the dissolved salts, this condition will be maintained so long as the cell is alive. The protoplasm, however, is enclosed by a firm cell-wall which prevents its indefinite expansion, so that the pressure generated within the protoplasmic envelope presses it closely against the cell-wall. The condition resulting from this pressure is spoken of as *turgor*, and cells in this condition are said to be *turgid*.

It is characteristic of turgid cells that the protoplasm is difficult to distinguish from the cell-wall, owing to being pressed closely against it, and that the cells are imbued with considerable rigidity on account of the internal pressure,—one may compare the difference in rigidity shown by a toy balloon or bicycle tyre when inflated and uninflated. The

---

* It is particularly the surfaces of the protoplasm which exhibit the characteristics of a semi-permeable membrane. In passing through the protoplasm which lines the cell-wall, therefore, a substance has to pass through *two* semi-permeable layers formed by the inner and the outer surfaces of the protoplasm.

rigidity of soft tissues of plants is due largely to the turgor of the cells composing them, and this rigidity is lost if the turgor of the cells is sufficiently reduced.

One of the ways by which the pressure within the cells can be reduced is by surrounding them with a solution more concentrated than the cell-sap. Under such conditions, water passes from the weaker cell-sap to the stronger solution without, the cells lose their turgidity, and the tissues become flaccid in consequence. If a single cell so treated is examined

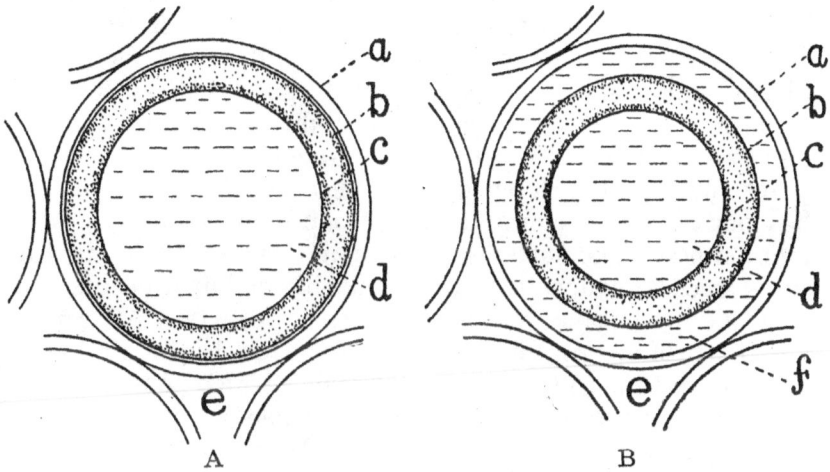

FIG. 4—DIAGRAM TO ILLUSTRATE CONDITION OF A LIVING PLANT CELL
A, turgid; B, after plasmolysis

a, cell-wall; *b*, outer surface of protoplasm; *c*, inner surface of protoplasm; *d*, vacuole containing cell-sap; *e*, intercellular space; *f*, plasmolysing solution after contraction of protoplasmic lining. In strict accuracy, the volume enclosed by the cell-wall should be smaller in B, since the wall is no longer stretched by internal pressure

microscopically, it will be found that (provided the surrounding solution is sufficiently concentrated) the volume of cell-sap is so much reduced that the protoplasmic sac is no longer large enough to fill completely the cavity enclosed by the cell-wall, and so becomes easily visible, especially if the cell-sap happens to be coloured. This condition of the cell, in which the protoplasm is contracted away from the cell-wall, is called *plasmolysis* and the cell is said to be *plasmolysed*. A plasmolysed cell can take up water and again become turgid, unless the previous treatment has injured the protoplasm.

Another way in which pressure within the cells can be reduced is by removing water, not by osmosis, but by simple evaporation. This occurs in the case of individual cells when an herbaceous shoot is left out of water, and becomes limp in consequence (p. 51). Yet another way is by killing the living protoplasm by means of heat, poisons, etc. Protoplasm has the properties of a semi-permeable membrane only while alive; if killed, it is no longer impermeable to the substances dissolved in the cell-sap, so that, with the escape of the latter, the pressure within the cell quickly falls. A cell which has lost its turgor in consequence of the death of the protoplasm cannot, of course, recover this turgor as can a cell which is plasmolysed, since the semi-permeability of the protoplasmic membrane has been destroyed permanently.

An important difference must be noted between the behaviour of protoplasm and that of non-living semi-permeable membranes. The latter differ among themselves as to the substances to which they are permeable or impermeable, but any one membrane is constant in its properties. The permeability of a protoplasmic membrane, on the other hand, alters from time to time, and a substance which passes readily through the protoplasm into the vacuole at one moment, may be unable to do so at another. It is in consequence of this varying permeability that the cell appears to exercise a power of selection over the substances entering and leaving it.

Before concluding this subject of permeability a few words may be said regarding the properties of the cell-wall in this respect.

The cellulose wall may be regarded as ordinarily permeable to all true solutions, but is impermeable to some *colloidal* solutions.* In respect to the latter, therefore, the cell-wall exhibits the properties of a semi-permeable membrane. Substances like sugar when in solution do not pass so readily through the cell-wall as does the water in which they are dissolved, although the wall is permeable to both. Consequently, if a solution of sugar is enclosed by a cellulose or parchment membrane and surrounded by water, the water will pass into the sugar solution more rapidly than the sugar will pass out. A temporary pressure is therefore produced within the

* For meaning of this term see p. 246.

membrane, gradually decreasing as the sugar slowly passes outwards, and ultimately disappears.

In considering the living cell, the behaviour of the cellulose membrane is evidently of minor importance as compared with that of the protoplasm.

Practical Work

1. Plasmodia of various Myxomycetes may be found on dead wood, fallen leaves, etc., in damp weather. If brought into the laboratory and at once placed under suitable conditions, the creeping movements described may be observed.

Place the material bearing the plasmodium in a vessel containing a little water, and arrange glass slides in sloping positions, so that their lower ends are in contact with the plasmodium. Arrange for a slow stream or drip of water to run down the slides by means of strands of wick or wool dipping into a vessel of water placed at a higher level than that containing the slides. Keep in a cool place, in a dim light, and observe at short intervals. The plasmodium and sporangia of a common Myxomycete, *Fuligo septica*, often may be found in tan-yards where oak bark is used; it is commonly known as " flowers of tan."

2. For microscopic examination of plant cells, small pieces of epidermis may be stripped off a leaf of *Iris*, *Narcissus*, or Hyacinth and mounted in water. *Chloroplasts* may be observed in small pieces of the leaves of water-plants or in the leaves of Mosses. *Nuclei* may be observed in almost any thin-walled cells, e.g. the cells of the epidermis of the leaves mentioned above. They can be rendered more evident by treatment with iodine, which kills the cell and stains the proteins of the cytoplasm and nucleus yellow. Good material for this purpose will be found also in the root-cap cells of the roots of seedlings of Oat or other plants. These cells separate easily from the rest of the tissues if the root tips are mounted whole in water (p. 61).

3. *Movements of protoplasm* in plant cells can be observed in the leaf-cells of the Canadian Pond-weed (*Elodea canadensis*), or in the purple hairs from the filaments of the stamens of the Spiderwort (*Tradescantia virginica*). In the former, the movement is a simple rotation around the cells, in which the chloroplasts are swept along; in the latter, each cell shows a complicated set of movements, due to the streaming of protoplasm towards and away from the neighbourhood of the nucleus.

4. Experiments to illustrate osmotic phenomena may be performed by using a piece of parchment tied over the broad end of a thistle funnel. Parchment is not entirely impermeable to sugar in solution, but if a strong solution of sugar is placed within the thistle funnel, the sugar escapes sufficiently slowly for a considerable rise of pressure to be produced, as is indicated by a rise in level of the liquid in the tube.

5. For microscopic investigation of *plasmolysis*, cells with coloured cell-sap are convenient, e.g. those of the staminal hairs of the Spiderwort, epidermal cells from the petals of *Viola*, thin sections of Beetroot, etc. The material should be mounted in water, 5 per cent. salt solution run in to cause plasmolysis, and recovery produced by irrigating with water.

6. The osmotic strengths of solutions may be compared by using a split Dandelion scape. The scape is split longitudinally into four pieces and one of the pieces thrown into a dish containing a solution of known strength—e.g. various dilutions of a gram molecular solution (see footnote, p. 19). It will be found to curl up or uncurl to a definite extent. When it ceases to show any further curling movement, it is transferred to the solution of unknown strength. The piece of tissue uncurls, curls up more, or remains unaltered according as the solution is stronger, weaker, or equal in strength to the known solution. The curling movement is produced because the outer tissue of the strip consists of cells with thick, inextensible walls (epidermis), while the inner part is composed chiefly of cells with thin, extensible walls (cortex).

7. Take a piece of turgid herbaceous tissue (e.g. scape of Dandelion, *Primula*, etc. or a young leaf of *Narcissus*) and measure its length. Soak pieces for some time in a 10 per cent. salt solution; they become flaccid and contract. Replace in water and remeasure after some time.

8. Take a similar piece of tissue, of measured length, and place for a few moments in boiling water. Note the permanent loss of turgor and contraction.

9. Take a piece of tissue with coloured cell-sap such as a petal of the flower of *Viola*. Slowly heat in a vessel of water containing a thermometer. Note the temperature at which the cell-sap escapes, thus indicating the death of the protoplasm. Leaves of species of *Oxalis* are particularly useful for this experiment; on death of the protoplasm, the acid cell-sap escapes and encounters the chlorophyll which thereupon turns brown and so causes immediate " browning " of the leaf.

CHAPTER II

RESPIRATION

THE PLANT AS MACHINE. FOOD AS A SOURCE OF ENERGY. UTILISA-
TION OF RESERVES. RESPIRATION OF SUGAR. AEROBIC AND ANAEROBIC
RESPIRATION. FERMENTATION. COMPARISON OF RESPIRATION IN
ANIMALS AND PLANTS. THE RESPIRATORY QUOTIENT. SEEDLING
NUTRITION

The Plant as a Machine—From the preceding chapter we
have learnt that a Flowering Plant is built up of parts or units
called cells. A machine also is built up of parts, but these
parts are not comparable with the cells of a living organism;
in fact, the comparison between a living plant and a machine
breaks down in this particular.

The component parts of a machine are made separately
and then brought together and " assembled." The complete
machine which results is capable of accomplishing work if
supplied with the necessary energy, ultimately derived from
some form of fuel; each individual part alone can accomplish
nothing.

The cells of a living organism, such as a Flowering Plant,
are not produced separately, but are derived one from another
by a process of cell division, growth and subsequent differentiation, as will be described in a later chapter. Each living cell
can manifest most of the activities regarded as characteristic
of life: *potentially*, each is complete in itself, *actually*, certain
forms of vital expression have been sacrificed. In contributing
to the complex individuality of the many-celled whole, the
identity of each cell is to some extent lost.

Just as a number of persons living together in a community find it is more convenient and economical for each to
specialise in a particular kind of work than for everyone to
be a Jack-of-all-trades, so in the higher plants and animals
we find that groups of cells are specialised to perform
particular kinds of work. In unicellular plants or animals,

25

in which one cell has to carry out all the functions of life, we find no specialisation and, in consequence, no restriction of activities. The phenomenon of specialisation in one direction with loss of capacity in others is found in the cells of all plants and animals except the very simplest.

In order to serve the interests of the organism as a whole, the activities of the specialised cells, or groups of cells, must be controlled and co-ordinated. The controlling force manifests itself in many different ways. Thus, the consumption of the reserve food of a plant is controlled in accordance with the needs of the plant at the time; leaves grow at definite places on stems, and the flowers of a plant are produced in a fixed and definite manner; removal of the main bud of a shoot results in a lateral shoot assuming its place. Of the real nature of the interrelation, or *correlation,* as it is called, between the different parts of a plant we know very little: it is, indeed, one of the chief mysteries connected with life. Occasionally this correlation is temporarily lost, and various abnormalities of structure and behaviour result. The cause for the loss of control is usually as mysterious as the nature of the control itself.

Food as a Source of Energy—We will now proceed to inquire as to the kind of food used by plants and to learn to what extent the way in which a plant makes use of its food materials may be compared with the way in which an engine uses fuel. In both cases energy changes occur, in the course of which the food or fuel is consumed. These inquiries will be best pursued, in the first instance, by means of observations and experiments upon germinating seeds and growing seedlings.

Seeds such as those of Bean, Pea and Wheat are valuable to human beings and other animals as food on account of the reserve food materials which they contain. The parent plant provides each seed with stores of food as a legacy upon which to start its career, or, in terms of our simile, with a store of fuel to enable it to carry on its life activities until such time as it is capable of obtaining fuel or food for itself. Such reserves occur in various chemical forms, one of the commonest of which is *starch.*

It should therefore be possible to observe that germination of the seed and growth of the seedling result in a decrease in the quantity of reserve food stored in the seed-leaves or cotyledons of the embryo plant, or in the endosperm of the seed; also that the growth of the seedling is checked if the store of reserve food be removed and no external source of supply can be drawn upon. Very simple experiments show that both these results take place (pp. 38, 42).

Starch belongs to the class of chemical substances known as carbohydrates, composed of the elements carbon, hydrogen and oxygen, the hydrogen and oxygen being present in the same relative proportions as in water, namely 2:1.

Since starch disappears from the storage regions of seeds during germination and growth is checked if this store of starch be artificially removed, one may deduce the fact that starch is one of the substances used as fuel by plants.

The starch consumed by the plant is put to two main uses: (1) to build new plant substance, and (2) to provide energy to carry on the life processes of the plant. Since the proto- plasm and cell-walls, which constitute the bulk of a plant apart from the cell-sap, contain the elements carbon, hydrogen and oxygen in high proportion, it follows that starch can contribute some of the main materials for constructing new plant substance. A large part of the starch or other carbo- hydrates comprising the food reserves is used, like the coal of a steam engine, for the purpose of providing energy to drive the machinery.

It is a well established physical principle, known as the " conservation of energy," that energy cannot be created or destroyed but only converted from one form to another. If therefore a change is to take place requiring the consumption of energy, an equivalent amount must be liberated from some other process. Thus, a steam engine is a device for con- verting the heat energy derived from the combustion of coal into the pressure energy of the steam, which in turn is con- verted into light energy, energy of movement, or any other form of energy that may be required, although it is never possible to convert the whole of the energy for disposal into the one single form that may be desirable at the moment. For instance, if we desire to use the energy available from an electric current for lighting a room, i.e. for producing light energy, we shall find that however efficient the lamps em- ployed, the light energy obtained will be less than the energy represented by the electric current consumed. This does not mean that any energy has been lost in the conversion, but that the balance has appeared in some other form, for example, as heat. The sum total of the various kinds of energy produced will be always equal to the amount of electric energy that has disappeared.

Many of the changes which occur in a growing seedling require energy for their accomplishment: the roots have to force their way into the ground, water has to be lifted up the stem, and chemical substances have to be produced which entail the consumption of energy.

Now the energy of a steam engine is derived from a chemical change in the fuel, namely, its combination with the oxygen of the air,—its combustion, as we say. Similarly, the energy required by a plant for the carrying out of its life activities is derived from a chemical change in its " fuel " or food, such being effected in the course of the process known as *respiration*.

The reserve food is stored in a special storage region of the seed: in the cotyledons of the embryo in a Pea seed, in the endosperm in a Wheat grain. This reserve food is not necessarily respired and the resulting energy liberated in these storage regions: every living cell in the plant respires in order to obtain energy for its life processes and consequently requires a share of the fuel or respirable food materials. All the cells, however, do not exhibit the same degree of vital activity at any one time: those in the growing regions, such as root tip or stem tip, show greatest activity. Actively growing regions, therefore, consume most fuel, i.e. respire most vigorously and have the greatest need for food to be sent to them. Young seedlings are growing very actively and therefore respire rapidly. They therefore make heavy demands upon food materials for growth and respiration.

Although starch is the form in which reserve food is often stored, it is never used directly as such but is changed into sugars of various kinds (sugars being also carbohydrates). The reason for this transformation is simple. Plant cells, as we have learnt, are completely enclosed by solid cellulose walls; starch cannot pass from one cell to another since it is in the form of solid grains which are insoluble in water or the watery solution we call cell-sap. In order that the starch may be made use of as fuel, it is transformed into sugars which are soluble in water, and can diffuse readily through the cell-walls of the storage cells to the parts of the plant where they are to be utilised. The advantage of starch over sugar as a reserve material depends upon its insolubility in water: for, being solid, it can be more economically stored in a limited space, and the troubles that would result from the high osmotic pressure due to the presence of an equivalent amount of sugar in the cell are avoided. The reserve food, thus converted into a soluble form, can travel from the storage region and, passing from cell to cell, be distributed throughout the living cells of the plant. In a Flowering Plant the transference, which may be over a considerable distance in the mature plant, is facilitated by a distributing system consisting of long tubes (sieve tubes) occurring in the part of the vascular system known as the bast or phloem.

Since the whole of the life activities are dependent upon the energy derived from respiration, this process is evidently of such importance as to merit a detailed inquiry as to its exact character. This inquiry we shall now proceed to make.

Respiration may be defined broadly as the breaking down of various organic substances in the plant with the result that energy is liberated and becomes available for the life-processes. The chemical changes involved are varied, but the most important in plants are those concerned with the oxidation of compounds of carbon, and the term " respiration " is often restricted to describe changes in carbon compounds such as result in the using up of oxygen and the production of carbon dioxide and water. In this restricted sense, respiration is essentially of the same nature as combustion, and differs from what happens when similar substances are burnt in the ordinary way, only in so far as it is initiated without raising the temperature.

Aerobic and Anaerobic Respiration—Let us consider this process when the material respired is grape sugar or glucose, using the term " respiration " to include changes which take place either in the presence or in the absence of oxygen. In the *absence* of oxygen, sugar breaks down into carbon dioxide and alcohol, energy being liberated in the process. Alcohol, however, as is well known, is capable of liberating energy on combustion. It is, therefore, not surprising that when sugar is respired in the *presence* of oxygen alcohol is not produced. The reaction proceeds as though the alcohol as soon as formed were oxidised to carbon dioxide and water, —additional energy being thereby liberated. The two cases may be tabulated thus:—

(1) In absence of oxygen,

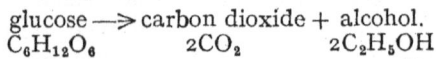

$$\text{glucose} \longrightarrow \text{carbon dioxide} + \text{alcohol.}$$
$$C_6H_{12}O_6 \qquad\qquad 2CO_2 \qquad\qquad 2C_2H_5OH$$

(2) In presence of oxygen,

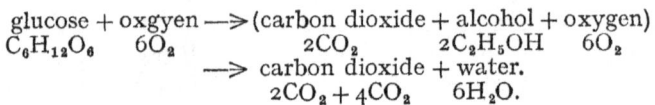

$$\text{glucose} + \text{oxgyen} \longrightarrow (\text{carbon dioxide} + \text{alcohol} + \text{oxygen})$$
$$C_6H_{12}O_6 \quad 6O_2 \qquad\qquad 2CO_2 \qquad 2C_2H_5OH \quad 6O_2$$
$$\longrightarrow \text{carbon dioxide} + \text{water.}$$
$$2CO_2 + 4CO_2 \qquad 6H_2O.$$

This statement is inaccurate to the extent that it is doubtful if alcohol is formed at all in the second case. In both cases, probably as a first respiration change of the glucose, some unstable, intermediate substance is produced: in the absence of oxygen this gives rise to alcohol; in the presence of oxygen it becomes oxidised to carbon dioxide and water. Whatever may be the intermediate steps, certain important facts should be noted in regard to the final products. In the first type of respiration, alcohol is produced, and this substance, in any but small quantities, is a poison to the living protoplasm. An organism which pursues this type of respiration must be able to get rid of the alcohol produced, otherwise this will gradually increase in concentration in the cell until it brings the vital activities of the organism to an end by its poisonous action.

In the second type of respiration there is no danger that, as the reaction proceeds, the vital activities will be checked in consequence of an accumulation of alcohol, since none is produced; carbon dioxide, being a gas, readily escapes. Furthermore, it is to be noted that a given quantity of sugar liberates a considerably greater quantity of energy than in the former case, the additional amount of energy being equal to what would be derived from the oxidation of the alcohol produced in the absence of oxygen.

Summing up, the second type of respiration differs from the first in three main particulars: (1) a supply of oxygen must be available, (2) the reaction does not lead to the formation of poisonous products such as alcohol, but only to that of carbon dioxide and water, (3) a much greater amount of energy is liberated from a given quantity of sugar.

Both types of respiration occur in plants: the first, carried on under conditions where oxygen is not available, is described as *anaerobic* (*aër*, air: βίος, life) and organisms which function in this way are described as *anaerobes :* the second, requiring a supply of oxygen, is termed *aerobic* and is characteristic of all animals and plants known as *aerobes.*

The higher plants resemble most animals in being typical aerobes: their cells can live for a short time if deprived of oxygen, but are quickly killed by the poisonous substances produced during anaerobic respiration. A number of the lower plants are typical anaerobes,—many species of bacteria, for example, and a number of fungi. Some of these plants are *obligate* anaerobes; that is, they are unable to exist under conditions in which oxygen is available; others are *facultative* anaerobes and can live and grow equally well either in the presence of oxygen or if deprived of this gas.

A suggestive speculation as to the mutual relations of these two forms of respiration is as follows. In the stages of the earth's history immediately preceding the appearance of life, free oxygen probably did not exist, or occurred in very small amounts: any oxygen present would have been used up in combining with various oxidisable substances. Under such conditions primitive organisms perforce must have been anaerobic; but they were also, presumably, unicellular, and poisonous by-products such as alcohol can easily diffuse away from isolated living cells. At a later stage of evolution, multicellular organisms appeared, and the removal of poisonous by-products from the inner tissues of a multicellular organism is attended with some difficulty. By this time, however, owing to the activities of chlorophyll-containing plants, considerable amounts of gaseous oxygen were available, aerobic respiration became possible, and the production or accumulation of poisonous substances in the tissues consequently was avoided. Respiration, in so far as it is a means of obtaining energy for the life processes, can take place in the absence of oxygen. It is possible that oxidation became a feature of respiration, in the first instance, as a means of destroying the poisonous substances formed during anaerobic respiration, the removal of such products becoming more difficult with the increase in size and complexity of the organism. That a much larger amount of energy was thereby rendered available, though of the greatest importance, was a secondary consequence.*

Having cleared the way with these general considerations, let us now turn to some actual cases of both kinds of respiration. Many aerobic plants have some capacity for living as anaerobes. If kept under conditions in which oxygen is available, aerobic respiration may be expected to predominate, since it renders available so much more energy from a given quantity of material. If we desire to investigate anaerobic respiration we will have to choose a unicellular plant living under conditions in which the oxygen supply is limited. Such is conveniently to hand in the Yeast plant, a facultative anaerobe. Let us take a little living Yeast and shake it up in a closed vessel which contains a weak solution of a sugar such as glucose. If allowed to stand in a warm place, the liquid will soon begin to froth owing to the evolution of gas. This gas may be demonstrated to be carbon dioxide. After some time the frothing ceases, and if the liquid be examined, three changes will be found to have occurred: (1) the amount of Yeast has increased, (2) the sugar has disappeared or decreased in amount, (3) alcohol has been produced.

This process, as it occurs in Yeast, whereby sugar is used up from the surrounding liquid and alcohol and carbon dioxide produced, has been known for a long time as *fermentation*. There is no doubt, however, that we are justified in regarding fermentation as a special case of anaerobic respiration.

Plants like Yeast can thrive without oxygen, provided the surrounding solution of alcohol does not become too concentrated. This is due to the ready diffusion of alcohol from the cells into the liquid without, and to the fact that Yeast is remarkably tolerant of alcohol, the concentration of which may rise to 10-18 per cent. without injuring the plant. There is an incidental advantage arising out of this power shown by the Yeast plant to withstand relatively large amounts of alcohol. Since alcohol is poisonous, even in small quantities,

---

* As a general rule the harmfulness of poisons to protoplasm is reduced as a result of their oxidation.

to most living cells, Yeast can evidently grow under conditions unfavourable to possible competitors.

For an example of aerobic respiration we will now return to our pea seedlings which make use of starch as a reserve material. Using these it is comparatively a simple matter to demonstrate that oxygen is taken from the air and an equal volume of carbon dioxide given off during their respiration. The amount of carbon dioxide given off compared with that of the oxygen used up is often expressed as a fraction, and is called the " respiratory quotient." In the present case the value of the respiratory quotient is evidently unity, thus:—

$$\frac{\text{Carbon dioxide}}{\text{oxygen}} \quad \text{or} \quad \frac{CO_2}{O_2} = \frac{1}{1}.$$

The value depends upon the substances respired and the nature of the chemical changes involved (p. 36).

Although pea seeds normally respire in this aerobic manner, giving out a volume of carbon dioxide equal to that of the oxygen absorbed, they have the power, to some extent, of respiring anaerobically. To demonstrate this fact, some germinating seeds are kept under conditions in which oxygen is deficient. In these circumstances carbon dioxide is given off, whilst the seeds acquire a distinctly alcoholic odour.

Plants differ among themselves as to the length of time they can withstand lack of oxygen. The higher plants being, like most animals, aerobic in their manner of life, soon succumb if they are unable to obtain oxygen. This is because they have no means of ridding their tissues of the poisonous products of anaerobic respiration. Pea seeds, for example, soon die when kept without oxygen, although, if deprived of oxygen for only a short time they will recover from the harmful effects, since the poisonous products of anaerobic respiration become oxidised and removed in the form, for the most part, of carbon dioxide and water.

Aeration of the more deeply seated cells is provided for as follows. The individual cells of a Flowering Plant do not fit closely together, but separate from one another at the corners where three or more cells meet. These *intercellular spaces* form a connected system of air channels through which oxygen gains access to the inner cells of the plant. Air, containing oxygen, not only diffuses into the deeply seated cells through this system of air spaces, but the gases evolved during respiration find thereby a means of diffusing away. In the case of water plants, where fresh supplies of air are difficult to obtain, this aerating system becomes much enlarged and forms a conspicuous feature of the internal anatomy.

The ordinary Flowering Plant is a typical aerobe, the Yeast plant may be taken as an example of a facultative anaerobe. Lastly there are obligate anaerobes, which can only thrive in the absence of free oxygen; examples of such are certain bacteria which live in the soil, and also others which are responsible for putrefactive changes in organic substances (p. 96 *et seq.*).

When sugar is respired, as when coal is burnt, some of the energy liberated appears in the form of heat. In plants, the rate of respiration is not usually so rapid as to cause their temperature to rise above that of the surrounding air. The heat is conducted away or otherwise dissipated almost as quickly as it is produced. If, however, rapidly respiring structures, such as germinating seeds, are kept under conditions in which heat cannot readily escape, e.g. in a Thermos flask, a rise of temperature can be demonstrated. Exceptionally, respiration is sufficiently intense in plants to cause a rise in temperature which can be recorded without taking these precautions, as in the well-known case of the opening inflorescence of Cuckoo-pint (*Arum maculatum*). This production of heat must be regarded, in general, as waste so far as the plant is concerned, since plants always assume approximately the temperature of their surroundings and can carry on their life activities within wide ranges of temperature.

This behaviour is in marked contrast with that shown by warm-blooded animals whose life activities are organised to function satisfactorily at a fixed temperature, a slight departure from which throws the vital machinery seriously out of gear. In this case the production of heat from the chemical changes accompanying respiration is not incidental but a necessity. A large amount of energy is therefore used in keeping the body of the animal up to the required temperature, especially when living in cold surroundings.

Utilisation of Reserves—The reader may have noticed that an important point has been left unexplained. We have spoken of sugar breaking down into carbon dioxide and alcohol, or of its oxidation to carbon dioxide and water. Now, at normal temperatures, sugar does not suffer either of these changes as the result of simple chemical reactions. How then do they occur in the plant cell ? The answer to this question will be found in the chapter dealing with enzymes (Chapter VIII.): for the present it is sufficient to note that changes can take place in living cells which can only be effected with extreme difficulty or at high temperatures in the laboratory.

We have considered the changes which befall starch when this substance forms a reserve food of the seedling. Starch, however, is not the only form of reserve material found in seeds. Thus, the reserve in Date seed is a kind of cellulose, a carbohydrate closely allied to starch in chemical composition, although differing from it in physical properties; in Rape and Sunflower seeds oils occur, substances which are composed of the elements carbon, hydrogen and oxygen, but are not carbohydrates, since the proportion of oxygen is less than in this class of bodies: nitrogenous reserve materials in the form of proteins, which contain nitrogen in addition to the above mentioned three elements, are also present in the storage regions of seeds, sometimes in considerable quantity. This nitrogenous reserve is stored in the form of granules called *aleurone grains*. These may be distributed throughout the cells of the reserve tissues, as in the cotyledons of pea or bean, where they occur as small, simple granules intermingled with the starch grains; or they may be confined to certain cells, as in the seeds of the cereals, where they are restricted to a layer immediately beneath the seed-coat, called the *aleurone layer*, from which starch is absent. In some seeds, especially those in which oil replaces starch as a reserve material, such as Castor Oil or Brazil Nut, the aleurone grains are large and of complex structure (fig. 3).

When these reserves are utilised, corresponding changes take place, as when starch is converted into sugar; i.e. in each case the insoluble reserve food is altered into some soluble substance which then travels in solution whither it is required. The mechanism by which these changes are effected is discussed later (Chapter VIII.).

The value of the respiratory quotient will obviously depend

on the nature of the reserve material respired and the amount of oxygen required to bring about complete oxidation. Thus, the value of this fraction in the case of starch is 1, but where oil is the respired material it has a value of less than unity, because the volume of oxygen taken in during respiration is greater than that of the carbon dioxide given out, since oil contains proportionally less oxygen than does starch, and therefore requires a larger supply to bring about complete oxidation.

Sugars are not found as reserve food in resting seeds, probably for the reasons given on page 29, but they are frequently encountered as secondary, soluble products derived from insoluble carbohydrates and are, in fact, the form in which the reserve food is most frequently respired.

Respiration in Plants and Animals Compared—It will be instructive to devote a few words to a comparison of respiration as it occurs in plants with respiration in animals. Like plants, animals obtain the energy they require for their life processes by means of chemical changes taking place in the cells of the body. With few exceptions these changes are of a nature corresponding to those occurring in aerobic plants and involve oxidation of the food or fuel with the production of carbon dioxide and water. Instead of oxygen reaching every cell of the body in gaseous form, however, there is a special system of blood circulation. Oxygen is absorbed by the red corpuscles of the blood when it enters the lungs, forming an easily dissociated chemical compound with the red pigment (hæmoglobin) of the blood. This red, oxygen-carrying blood is then pumped by the heart to all parts of the body. The living cells remove oxygen as they require it from the blood stream, causing the red arterial blood to change in colour, while the products of the chemical actions taking place in the cells are absorbed by the blood and carried away. Of these products, the carbon dioxide and some of the water are given off from the blood when it again passes through the lungs and are expired in the breath. Although there is nothing in plants to correspond to the circulation of the blood as a means of providing the cells with oxygen, the actual chemical changes by which energy is made available are very similar except in one particular.

In seedling plants, the substances used up during

respiration are notably those containing the elements carbon, hydrogen and oxygen, and the same is true for the mature plant. In the chemical changes accompanying respiration in animals, however, compounds containing nitrogen, belonging to the class of substances known as proteins also play a conspicuous part. This is evidenced by the provision, in the higher animals, of a special system in the kidneys for freeing the blood of nitrogen-containing " waste " derived from the use of proteins as " fuel " in the living cells.

That plants use proteins under normal conditions as fuel to provide energy is doubtful, although they may do so under starvation conditions. Substances derived from the oxidation of proteins are usually poisonous to living cells. The animal provides for their excretion by means of a special organ: the green plant is a more economical machine, and instead of excreting these substances if formed, builds them up once more into complex protein material. Thus, although it is possible that proteins may be involved directly or indirectly in the respiratory processes of plants, energy is not thereby rendered available for the general life processes, since any energy set free by the breaking down of proteins is consumed in their reconstruction. Apart from respiration, the need of the seedling for protein reserves is obvious enough, since growth and development demand the formation of new protoplasm, and protoplasm consists largely of proteins.

Seedling Nutrition—The nutrition of a young seedling, dependent on the food reserves of the seed, has evidently much in common with that of an animal, both as regards the nature of the food materials and the manner in which they are used. In both, organic food substances such as carbohydrates, oils and fats are used for the building up of new plant and animal tissues and also for the liberation of energy as the result of respiration. During respiration, carbon compounds are oxidised and decomposed in a manner comparable with the combustion of fuel in an engine.

In the animal and in the seedling respiration ordinarily involves the taking in of oxygen and the production of carbon dioxide and water. When coal is used as a fuel in an engine, similar chemical changes occur in the carbon compounds and the same end products are produced, although the comparison is obscured by the fact that coal is of more complex chemical composition than sugar and there is consequently a greater variety in the end products resulting from its combustion. Compounds of sulphur, for instance, are present in the smoke from coal fires and the ash left behind consists of mineral material. Some of the substances formed may be due to incomplete combustion, and these may be compared with the alcohol formed during anaerobic respiration, i.e. with the product of incomplete oxidation of sugar in the living cell.*

When fuel is burnt, it is possible to account for the whole of the original weight by collecting the products of combustion. Since, ordinarily, most of these are dissipated in smoke, we simply observe that the original solid bulk or weight of coal gradually decreases, and is finally represented only by the *ash* and *soot*. Similarly, in spite of the fact that the *bulk* of a seedling is considerably greater than that of the seed from which it was grown, it can be shown by drying and weighing that the *dry weight* (i.e. the solid matter) of a seedling, grown under conditions such that the supply of food materials is limited to the reserves contained in the seed, is considerably less than that of the ungerminated seed. Since starch, sugar, and other materials are broken down into substances which are eliminated from the plant, respiration must involve a loss of material. The decrease in dry weight is accounted for by escape of the carbon dioxide and water formed during respiration. If these are collected and weighed, they will be found to equal the loss observed, if we add to it the weight of the oxygen used. The rate of loss varies in different plants and at different periods of growth, depending on the intensity of respiration. It is normally made good by the absorption of fresh food materials; if this is prevented the plant will gradually become depleted of respirable material and eventually die of starvation. By-products, corresponding to the ash or soot, are not formed because we are dealing with a substance, sugar, which can be completely oxidised to carbon dioxide and water (p. 30).

Incidentally, we learn that a large part of the increase

* It may be noted that, when heated in the absence of air (i.e. oxygen), coal breaks up into a large number of complex products, many of which are made use of in industry, instead of suffering the simple oxidation that takes place with a free air supply.

in size of a plant during growth must be due to the absorption of water.

When the seedling grows into an independent green plant it acquires new and remarkable powers and the nutrition differs in important particulars from that described for the seedling. The independent plant, like the young seedling, demands supplies of organic food for respiration and nutrition, but has become independent of external sources of supply. By reason of the activities of its chlorophyll-containing cells, it becomes, in fact, a self-supporting organism which can manufacture the necessary food from raw materials such as carbon dioxide, water and mineral salts.

Organisms like the animal and the young plant, which depend upon an external supply of organic food material to live and grow, are described as *heterotrophic* (ἕτερος, other; τροφή, food) in their nutrition: those which, like the mature green plant, can utilise simple inorganic substances from which to build up the necessary food materials are said to be *holotrophic* or *autotrophic*.

We will now pass on to consider the nutrition of the green plant in greater detail.

Practical Work

1. Set up an experiment as follows to show the " fermentation " of sugar by Yeast.

Place a little living Yeast in a 5 per cent. solution of cane sugar in a flask (fig. 5). Connect the flask by means of a bent tube with a vessel containing lime-water, and leave in a warm place. After a time, note the evolution of bubbles in the flask. By drawing a current of air through the apparatus the gas (CO_2) which is formed passes over into the vessel, and, as it bubbles through the lime water, forms a white precipitate of calcium carbonate (chalk). The presence of alcohol in the flask can be detected by the smell. In the apparatus shown, the air entering the flask is deprived of CO_2 by passing through a tube *b* filled with soda-lime.

N.B. The Yeast can be grown either under *aerobic* or *anaerobic* conditions by leaving the flask open to the air, or fitting it with a cork as shown in the diagram.

Note the effect on the accumulation of alcohol in each case.

2. That germinating seeds require oxygen is well shown by the following simple experiment.

Place a few Barley grains in each of two vessels A and B (fig. 6). Cover the grains with a funnel in each case. Fill the

vessels with water and arrange A so that water from a tap drips
into the tube of the funnel. The effect of the arrangement is
that the seeds in A are given an adequate supply of oxygen by

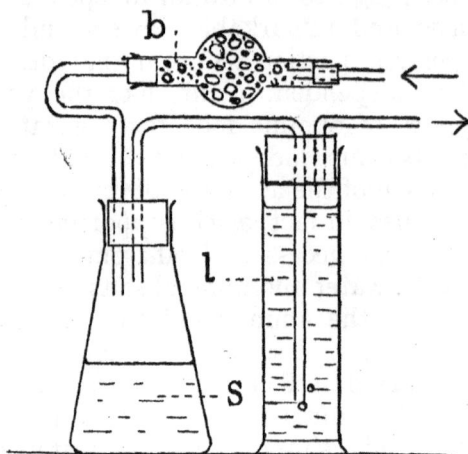

FIG. 5—APPARATUS TO SHOW THE EVOLUTION OF CARBON DIOXIDE
BY YEAST IN A SOLUTION CONTAINING SUGAR

b, tube containing soda-lime to absorb carbon dioxide from the air
    entering the flask; *s*, sugar solution containing yeast; *l*, lime-
    water. (The left-hand tube may, with advantage, dip into the
    solution *s*.)

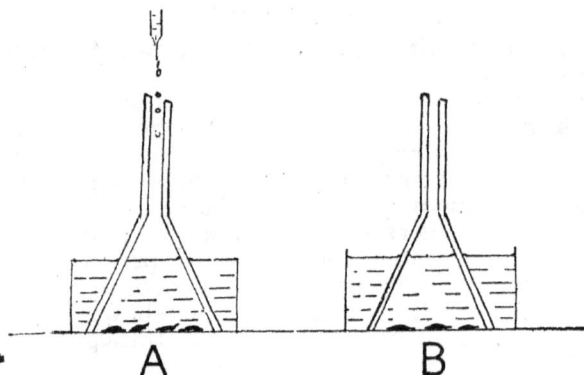

FIG. 6—EXPERIMENT TO SHOW THAT SEEDS REQUIRE OXYGEN IN
ORDER TO GERMINATE

A, seeds under water aerated by means of a stream of water; B, seeds
    in unaerated water

constant aeration of the water, whereas those in B are dependent
on the small amounts originally dissolved in the water or going
into solution from the air. It is instructive to set up a " control "

experiment, using similar seeds placed on moist blotting paper, and to compare the effect of the lower temperature experienced in the aerated water. Allowing for this, barley seeds germinate equally well under water if a supply of oxygen is ensured. The conditions in B approximate to those existing in badly-aerated or waterlogged soil.

Note—Seeds other than those of Barley can be used. The latter is recommended because it withstands a low temperature well and therefore serves for the experiment even when the temperature of the water is low.

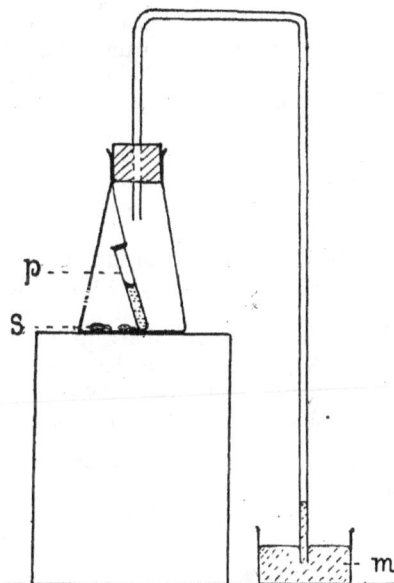

FIG. 7—APPARATUS FOR RESPIRATION EXPERIMENTS WITH GERMINATING SEEDS. (See text)

s, seed; *p*, tube containing potash (KOH*); *m*, mercury

3. The evolution of CO_2 by germinating seeds can be demonstrated by using the same apparatus as in experiment 1, placing in the flask a number of seeds of e.g. Pea instead of the sugar solution and Yeast.

4. **The Respiratory Quotient**—Set up the apparatus shown in fig. 7, placing in the flask one or two pea seeds on moist blotting paper. Cover the flask with a dark cloth, and note the position of the mercury in the tube when the experiment is started and at the end of a few days. Infer any change in the volume of the gases contained in the flask.

* Caustic soda (NaOH) may be substituted for caustic potash in this and other experiments involving the absorption of CO_2.

Set up a similar apparatus, suspending inside the flask a small glass tube containing caustic potash to absorb any carbon dioxide present. Note the subsequent rise of mercury in the tube in this case and infer the reason.

Precautions—The flasks should not be too small: they should be closed by rubber corks and all joints made absolutely air-tight. A few seeds only should be placed in each flask, owing to the small volume of air available. The flasks must be uncorked after a few days for the same reason if it is desired to prolong the experiments.

To test whether the flasks are air-tight. At the beginning of the experiment place the lower part of the flask in warm water for a few moments: remove, and quickly place the end of the tube under mercury. As the air in the flask cools, it contracts, and the mercury rises in the tube. The air-tightness or otherwise of the flask can then be judged by the behaviour of the mercury column. For more advanced experiments in respiration an apparatus must be designed which allows a current of air to pass through the chamber in which the seeds or other respiring organs are placed.

5. Loss of weight due to respiration can be shown as follows. Select two lots of pea seeds of (as nearly as possible) equal weights. Take one lot A; dry at 100° C. until they cease to lose weight. Weigh, and record the dry weight. Germinate those of lot B on moist blotting paper in a germinator in the dark. At the end of a fortnight, take the resulting seedlings, dry at 100° C. until they cease to lose weight, and compare the dry weight with that of the ungerminated seeds (A).

Infer the reason for the loss of dry weight. Compare experiment 6, Chapter V.

CHAPTER III

THE WATER RELATIONS OF PLANTS

THE WATER RELATIONS OF PLANTS. STRUCTURE OF ROOT AND SHOOT IN RELATION TO ABSORPTION OF WATER AND TRANSPIRATION. ABSORPTION OF WATER FROM SOIL: "AVAILABLE WATER." TRANSPIRATION: EFFECT OF EXTERNAL CONDITIONS ON TRANSPIRATION. HYDATHODES. THE "TRANSPIRATION CURRENT." FORCES CONCERNED IN THE ASCENT OF WATER

HAVING gained an insight into the manner in which a plant makes use of its " fuel " or food, we turn now to consider the nutrition of the plant in a more general sense,—how food is obtained from the outside world and the way in which it is used. In Flowering Plants there is much division of labour in the plant body and the work of nutrition is carried on by separate organs. Before we can proceed therefore, we must review broadly the architecture of the plant machine, in so far as a knowledge of this is necessary for an understanding of the facts of nutrition. Complete information concerning the anatomy of plants does not come within the scope of the present volume and should be obtained from a textbook of plant anatomy.

The Root System—The root system of a plant is in contact with a large volume of soil. Whether the soil so laid under contribution is in the superficial or deeper layers depends partly on the mode of branching of the root, but is greatly influenced by the position of the available water supply.

The growing portion of a root, whether of a main or a lateral root, is near the tip and includes two distinct regions: (1) the apical meristem in which new cells are produced, and (2) the " region of elongation " in which the new cells complete their development. The former is nearer the tip and gives rise to new plant tissues on both sides. That towards the tip goes to form the root-cap, that on the side away from the tip adds to the length of the root. As a result of the position of

43

the region of elongation, the root-apex is pushed through the soil and the root-cap serves to protect the growing region from injury in the process. The root-cap naturally suffers abrasion, hence the necessity for the continual replacement of its tissues.

If we examine the root of a seedling germinated in moist air, we notice a zone extending from near the tip backwards covered with fine hair-like structures,—the *root-hairs*. Microscopically, each root-hair is a tubular outgrowth from the superficial layer of the young root, hence called the hair-bearing or *piliferous layer*. Each hair starts growth as a bulge on the outer wall of a cell of the piliferous layer and grows out to form a slender tube. If the seedling is germinated in soil instead of in moist air, the root-hairs find their way between the soil particles to which they become so closely applied that it is impossible to detach all the soil from the root without damaging them. Owing to the formation of root-hairs, the surface area of the root capable of absorption is enormously increased, a significant fact since absorption from the soil is practically restricted to the root-hair region of the plant. In older parts an impervious layer or layers is formed on the outside of the root, either by the formation of cork, or by alteration of the walls of the cells beneath the piliferous layer. In either case, the water and food supply is cut off from all the tissues without, and the root-hairs wither and disappear. It should be noted that root-hairs grow out only from the region in which elongation has ceased to take place; they are not therefore damaged by movement of the root through the soil.

External examination of the root reveals the following regions: (1) The *root-cap* which receives the brunt of pushing through the soil, the loss it suffers from abrasion in this process being made good by the formation of fresh tissue from (2) the *growing region* just behind it. The growing region includes the *apical meristem* where new cells are formed and the region of *differentiation* and *elongation* behind it, and is thus responsible for the growth in length of the root as a whole. (3) The region bearing root-hairs which become older and longer as they recede from the tip of the root. (4) The older part of the root, furthest away from the root-tip, devoid of root-hairs and concerned only with conduction

The Shoot System—The shoot system of the plant may be aerial or subterranean. In either case, it consists of a branched or unbranched stem or axis bearing leaves.

In the case of subterranean shoots, the leaves are usually small and colourless or reduced to scales, and the shoot may superficially resemble a root. Such underground shoots may perform important functions, notably the storage of food materials and the reproduction of the plant by vegetative means; but the more characteristic and important functions of shoots compared with roots are carried out by the green foliage leaves. Elongation of the stem takes place chiefly at the apex, although *intercalary* growing regions occur in the case of some stems.

The apical meristem at the extreme tip of the stem gives rise to new cells which, as they are left behind by growth of the tip, differentiate in various ways to form the tissues of the mature stem. It gives rise, also, to leaves and to the lateral buds which arise in their axils. These lateral members of the stem,—leaves and lateral branches,—arise in regular order, as is shown by a vertical section through the apex of an ordinary shoot.

The regular development of leaves in this way is responsible for the formation of *buds*, each of which consists of the young growing apex of a stem overarched by a regular series of young leaves, or, in the case of the flower, by the young floral leaves. The region of elongation of the stem extends backwards from the tip, and it is here that cells grow to their full size and differentiate to form the tissues of the mature stem. As a consequence of elongation, the young leaves are carried apart and internodes are developed between them.

The whole of the shoot-system, including the stem and leaves, is covered with a (usually) single layer of cells called the *epidermis*. The chief function of the epidermis is to protect the plant from mechanical injury and to prevent loss of water by evaporation from the surface cells. To this end the external walls of the epidermal cells are thickened and their outer layers impregnated with *cutin*, a substance impervious to water. This impervious outer layer is called the *cuticle* and varies in degree of development in different plants and in different parts of the same plant. The cells of the epidermis frequently give rise to one-celled or many-celled hairs, which have an astonishing variety of form and function according to the species of plant, but are quite unlike the delicate root-hairs produced by the piliferous layer of the root. The epidermis is not continuous but is perforated by slits between certain pairs of cells. Each of these apertures, together with the cells immediately surrounding it, is called a *stoma* (pl. *stomata*); their structure and function are dealt with in the paragraph describing the structure of the leaf (p. 52).

The Conducting or Vascular System—Both root and stem are provided with an elaborate vascular system which forms a continuous network of conducting tissues throughout the plant, the character and arrangement of whose components depend on the species of plant and the part of the plant examined. This vascular bundle system consists of two main portions, the *xylem* or *wood* and the *phloem* or *bast*, both of which are concerned mainly with conduction. The conducting tissues of the wood are used to carry water and soluble substances to the leaves and other organs, those of the bast to convey away from the leaves the organic food materials manufactured there and to distribute them where needed. The diversity of arrangement of these vascular bundles found among plants is almost endless, but some arrangements are commoner than others and may be regarded as typical of the plant parts in which they occur. Thus, in young roots of Flowering Plants, it is usual for the strands of wood and bast to be arranged alternately in a single ring as seen in transverse section of the root; in young stems the wood and bast strands are united to form single bundles, the woody portions of which are directed towards the centre of the stem. The arrangement of these *collateral bundles* as revealed by transverse sections through the stem is either in a single ring as in the Dicotyledons, or scattered with apparent irregularity throughout the stem as in the Monocotyledons. In all cases, branches are given off to supply the lateral members,— lateral roots, lateral shoots, leaves and flowers,—and the strands may communicate with one another by cross-connections as at the nodes of the stem. A further complication arises in the case of roots and shoots of perennial Dicotyledons which undergo a process known as *secondary thickening*. Such roots and shoots increase continually in girth during the active life of the plant, owing to the addition of new layers of wood and bast. These additional tissues are formed in such a manner that the wood occupies the central portion of the stem or root and the bast forms a thin ring around it, the secondary wood and bast being separated from one another by the layer of cambium from which they have both arisen. Moreover, the outer tissues of such plants become replaced by a zone of cells with corky walls, which constitutes the major portion of the protective tissue known as *bark*. The manner in which these different tissues arise is indicated in Chapter IX.

The chief functions of roots are: (1) *absorption* of water and mineral salts from the soil; (2) *fixation* of the plant firmly in the soil, a matter of the greatest importance to plants, such as trees, with an upright shoot system; (3) *subsidiary functions*, such as anchoring the stem to its support, e.g. the climbing roots of ivy; storing up food material as in the swollen root of the dandelion; assimilating carbon as in the air-roots of some orchids, and so on.

The principal functions of the shoot are: (1) in the case of the stem, support of the leaves, flowers and fruits, and distribution of food materials throughout the plant; (2) *assimilation of carbon* from the carbon dioxide of the air by the green tissues; (3) *transpiration* or the giving off of water vapour which, as we shall learn, affects directly the rate of absorption by the roots; (4) *subsidiary functions*, as, for example, storage of food, vegetative reproduction, formation of special organs for climbing, etc.

We will commence our study of nutrition by a consideration of the part played by water in the life processes of plants.

The Water Relations of Plants—The proportion of water in plant tissues varies with the species and the particular part of the plant examined. One extreme is represented by the Algæ and the succulent parts of Flowering Plants in which water may constitute 95 parts in 100 parts of the fresh tissues; at the other end of the series are structures such as seeds, which, when air-dried, contain only from 10 per cent. to 15 per cent. of water. It must be remembered that active protoplasm contains a high proportion of water, withdrawal of which results in a dormant condition, as in seeds, or in death, as when a leaf is allowed to dry up.

Normally, a plant takes water from the soil by means of its roots, passes this up the stem for distribution all over the plant, and gives off large quantities in the form of vapour into the surrounding air, chiefly from the leaves. There are evidently three stages in the process; (1) absorption from the soil by the roots, (2) distribution throughout the plant, (3) escape of water vapour into the air.

ABSORPTION OF WATER

Absorption of Water from Soil—Soil consists of fine particles of varying nature, each of which, in a well drained soil, is covered with a film of water (see Chapter XV.). In close contact with these particles are the roots of plants growing in the soil. Extreme intimacy of contact is ensured in most plants

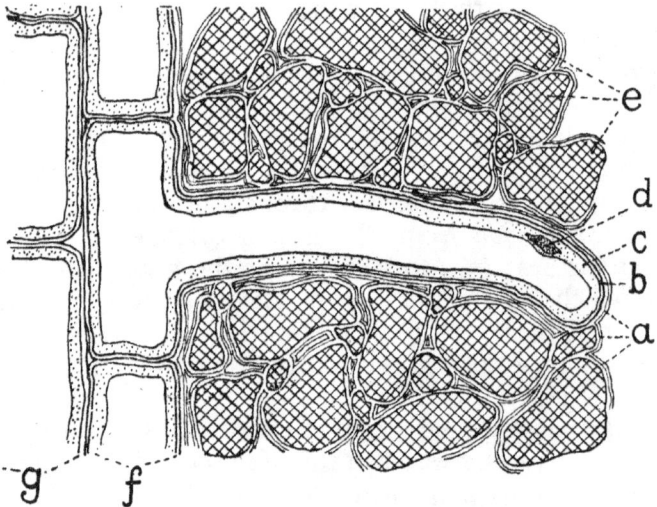

FIG. 8—DIAGRAM SHOWING THE RELATION BETWEEN ROOT-HAIRS AND SOIL PARTICLES

e, soil particles; a, films of water; b, wall of root-hair cell; c, cytoplasm; d, nucleus; f, cells of piliferous layer of root; g, cells of cortex of root

by the formation of root-hairs, which force their way between the soil particles and so render available the largest possible surface for absorption. Entry of water into the plant can be explained by reference to what has been learned of the osmotic properties of the cell (p. 17). The cell-sap of the root-hairs contains various osmotic substances, such as sugars and mineral salts, in solution; the soil water is an extremely weak solution of mineral salts. These two solutions of unlike concentration are separated from one another by the thin cellulose wall of the root-hair and its protoplasmic lining, the latter a semi-permeable membrane, permeable to water although impermeable to many of the substances in solution. Under these conditions, water passes from the weaker solution without through the semi-permeable layer of protoplasm into the root-hair. The capacity of the root-hairs to remove water from the soil is perhaps not surprising when we remember the large value of the osmotic pressures produced by even dilute solutions (p. 18).

Since the root-hairs absorb water by osmosis, it should be possible to check the process by surrounding them with a concentrated salt solution. A condition of this kind occurs in Nature in salt-water marshes and other soils in which the amount of soluble materials is abnormally high. Plants growing in such soils may find it difficult to absorb sufficient water for their needs; they may suffer from *physiological* dryness, although their roots are *physically* moist.

The absorption of water being largely an osmotic phenomenon which depends on the fact that living protoplasm behaves as a semi-permeable membrane towards many of the substances dissolved in the cell-sap, we might expect that conditions which affect the protoplasm directly would react upon the capacity of the root to absorb water from the soil. This expectation is justified by experiment since conditions which affect the vitality of the protoplasm adversely, such as lack of oxygen or a low temperature, reduce the capacity of roots to absorb water from the soil. It is to this cause that the injury suffered by vegetation from cold in temperate latitudes is often to be attributed. When the ground is chilled, roots may be unable to absorb water with sufficient rapidity to replace that lost by transpiration, especially when the condition is one of low temperature accompanied by drying winds. That plants under such conditions are often protected from injury by covering them with a bell-jar, straw, etc., is due to the means adopted checking rapid loss of water from the parts above ground, rather than to any protection afforded the plant against cold.

4

" **Available Water** " **in Soil**—It is important to distinguish between the *total water* and the *available water* in soil. The whole of the water in a soil is not available to the plants growing in it. Water is absorbed by the roots only so long as the osmotic forces tending to cause water to enter the root cells are greater than the forces acting in the contrary way. These latter are: (1) the osmotic pressure of the soil solution around the roots, (2) the force of surface tension tending to hold the water to the soil particles, (3) any other forces tending to hold water in the soil, such as those associated with the presence of a high proportion of colloids (p. 246).

Determination of the amount of available water is beset with difficulties. The usual method is to allow a pot-plant to use up the water from the soil in which it is growing until it begins to wilt, no fresh supplies of water being allowed to reach the soil. Wilting signifies that the water lost from the leaves is not compensated in amount by that taken up by the roots, and it is then inferred that the soil water available to the plant has been exhausted. The amount of water still retained can be estimated by drying a sample of soil at 100° C. and noting the loss of weight, but this water must be regarded as unavailable for the use of the plant. It is important to note that such an experiment indicates only that the amount of water in the soil at the moment when the experimental plant wilts is insufficient in amount to allow absorption by the roots to keep pace with the loss of water from the leaves, *under the conditions existing during the experiment*. Since both the rate of absorption by the roots and the rate of loss from the shoot are largely controlled by external conditions, it is evidently necessary to carry out such experiments under prescribed conditions, if the results are to have any comparative value.

Absorption of Water by Leaves—In the case of submerged water plants, absorption of water takes place by all parts of the plant, including the leaves. In terrestrial plants, absorption by the mature leaves is rendered practically impossible owing to the presence of the cuticle. Certain plants, growing under exceptional conditions, absorb water readily by means of their leaves which exhibit a special structure to this end.

LOSS OF WATER

Escape of Water Vapour into the Air: Transpiration—Any wet surface exposed to the air gives off water vapour by evaporation, unless the air is saturated with moisture. Consequently, an isolated plant cell tends to lose water by evaporation, since the cell-wall is kept moist by contact with the protoplasm surrounding the cell vacuole. In the case of unicellular plants, therefore, we should expect the cells to dry up rapidly and lose most of their water if exposed to the air. For this reason such plants usually grow in water or in situations where the air is saturated with moisture; or loss of water by evaporation is checked by investment of the cell with mucilage.

If living cells lose a high proportion of water, they may become dormant until a fresh supply is forthcoming, and are frequently none the worse for this suspension of their vital activities. This is the case, for example, in the cells of the embryo of the resting seed; certain of the lower plants, also, can suffer desiccation without injury, e.g. the leafy liverworts and the Lichens. Vegetative cells of the higher plants do not usually recover from such an experience, either because the protoplasmic threads connecting one cell with another shrink and break, or because the cell-sap becomes so concentrated as to injure the protoplasm.

Evaporation from the outer surface of a Flowering Plant is almost entirely prevented by the presence of the cuticle, the development of which is most conspicuous in plants growing in situations such as expose them to the danger of losing an excessive amount of water. Not infrequently this effect of the cuticle in checking loss of water from the surface of the plant is supplemented by a layer of wax.

There are, however, air passages or intercellular spaces throughout the living tissues, which serve for the circulation of gases. The cell-walls bordering on these air spaces are in no wise protected from loss of water by evaporation, for which reason the air within the spaces speedily becomes saturated with moisture. This system of air spaces is in communication with the outside air through the stomata which perforate the epidermis. Water is thus lost from the plant by evaporation from the walls of the cells which line the intercellular spaces, and thence, by diffusion of water vapour through the stomata, to the outer air (see fig. 15, p. 86).

The structure of a stoma is such that the size of the opening varies automatically with the needs of the plant, so controlling the rate of diffusion of water vapour from the intercellular spaces, and consequently the amount of water lost by the plant. Each stoma arises in the following way. In the young leaf all the cells of the epidermis are alike. As differentiation of the tissues takes place, certain pairs of cells behave differently from the rest. The walls that are in contact separate for part of their length, leaving a slit or passage between two cells known as the *guard cells*. The walls of the guard cells become locally thickened, especially in the neighbourhood of the opening; when turgid, the cells curve away from one another as a result of the internal pressure and the unequal extensibility of their cell-walls. If, for any reason, the guard cells become less turgid, they straighten by the elasticity of their cell-walls, so closing the opening between them. Thus, the stomata remain open so long as the plant has sufficient water and the cells are turgid, but they close automatically when the plant loses so much water that the cells cease to be turgid.

In reality, the matter is not quite so simple as described for two reasons. Firstly, the structure of the guard cells is more complex than has been described, and in the second place, the movements of these cells, and consequently the condition of the openings are controlled, not only by the water content of the plant, but also by other conditions such as illumination.

The stomata are distributed on all the young aerial parts of the plant, but are particularly characteristic of the leaves, the organs from which water vapour is chiefly given off. This loss of water from the plant by evaporation of water vapour into the intercellular spaces and its subsequent diffusion through the stomata is called *transpiration*.

Structurally, the leaf may be pictured as a loose network of thin-walled cells, forming a plate of tissue several cells in thickness, supported and supplied with water by a framework of woody tubes in the vascular bundles or veins of the leaf. Covering the whole leaf is the epidermis, perforated with numerous stomata which, by their power of opening and closing, can control the rate at which water vapour escapes by diffusion from the saturated atmosphere within. When the stomata are closed, a continuous waterproof covering, the *cuticle*, practically prevents loss of water from the interior of the leaf; when they are open, it is as though the epidermis were perforated with holes so numerous that water vapour can escape from the leaf almost as rapidly as if the epidermis were absent.

Experiments upon the rate at which diffusion takes place through plates perforated with small holes have thrown light upon the use of the stomata in allowing diffusion to take place between the inside of the leaf and the outside air. It is found that as the number of holes in a given area increases, the size of the holes becomes of decreasing importance; in other words, diffusion takes place very much more rapidly through a plate perforated with a large number of small holes than through a plate with a smaller number of large holes, even though the total area of the holes is the same in the two cases. The stomata of the epidermis, although small, are very numerous, and therefore allow water vapour and gases to diffuse backwards and forwards with little hindrance. It has been calculated that when they are open diffusion can go on between the internal atmosphere of the leaf and the outside air almost as rapidly as if the epidermis were not present. On the other hand, it must be remembered that if the stomata are closed practically no loss of water vapour takes place. By means of its stomata the leaf can thus control the rate of transpiration within wide limits.

The distribution of stomata on the leaf varies in different plants: in some, e.g. Broad Bean, they are present in equal numbers on both sides of the leaf; in a majority of plants, they are more numerous on the under side of the leaf, or are absent from the upper side; in floating leaves, such as those of Water Lilies, they are confined to the upper surface of the leaf and are absent from the lower, submerged side. The size of the pore when open varies in diameter from 0·007 milli-metre in Field Spurry (*Spergula arvense*) to 0·02 millimetre in some kinds of lily (*Amaryllis sp.*), the number occurring in one square centimetre of upper and lower leaf surface respectively being 0 and 800 in the former plant, and 9 and 28 in the latter. An average sized leaf of Sunflower possesses about 13 million stomata.

The amount of water vapour given off in transpiration is considerable. It has been calculated that a birch tree with about 200,000 leaves loses on an average $15\frac{1}{2}$ gallons of water in a day, while on a hot day it may lose as much as 85 gallons. The amount of water vapour given off by the whole of the vegetation in a district is thus very great, and profoundly affects the humidity of the air and consequently the climate. It is known, for instance, that felling large numbers of trees may react indirectly upon the vegetation of a district by decreasing the rainfall, owing to the diminished humidity. From this point of view, a covering of vegetation acts as a potent means whereby water is removed from the soil and given off as vapour into the air, whence it may once more be returned to the soil in the form of dew or rain.

Stomata have important functions other than that of controlling the rate of transpiration. The cuticle is impervious not only to water and water vapour, but also to gases such as oxygen and carbon dioxide. It is through the stomata, therefore, that all interchange of gases takes place, and through them that communication is established between the intercellular air passages of the leaf and the outside air.

To summarise what has been learned regarding transpiration. A Flowering Plant gives off water vapour or *transpires* from its aerial parts, especially from the leaves. Transpiration occurs in two stages: (1) water evaporates into the intercellular spaces, which become saturated with water vapour; (2) water vapour diffuses through the stomata into the outside air. The rate of diffusion and consequently the rate of transpiration depend largely upon the condition of the stomatal pores, the opening and closing of which are controlled by the movements of the guard cells. There is little doubt that the osmotic condition of the cells of the leaf surrounding the intercellular spaces must affect the ease with which water evaporates into the latter. It is not agreed at present to what extent such differences react directly upon the rate of transpiration.

Effect of External Conditions upon the Rate of Transpiration—Since the air within the intercellular spaces of the leaf is saturated with water vapour, the rate of transpiration is

largely determined by the action of those factors which control
the rate of diffusion from the intercellular spaces into the air
surrounding the plant. These factors are of two kinds:
those dependent on the structure of the plant itself, and those
due to the external conditions under which it is growing.
To the former category belong structural features of the plant
such as, the area of surface exposed to the air, the number of
stomata in a given area of leaf, and the thickness of the cuticle.
To the latter belong the physical conditions which affect
the rate of diffusion.   Provided that the plant has a plentiful
water supply and that the stomata are open, the rate of transpiration will be rapid or the reverse, according to whether the
conditions are such as favour the diffusion of water vapour
or not.   Thus, conditions tending to increase transpiration are
dryness of the air, and movements of the air whereby water
vapour is removed from the immediate neighbourhood of the
stomata.   A high temperature also favours a rapid rate of
transpiration, because diffusion occurs more rapidly at high
temperatures, and the amount of moisture required to produce
saturation of the air without is greater.

It is a remarkable illustration of the way in which the
structure of plants is adjusted or adapted to the conditions in
which they are growing, that plants growing in damp situations
usually possess large leaves with numerous stomata and a
thin cuticle, whilst those growing in situations *physically*
or *physiologically* dry have small leaves with few stomata
and a thick cuticle.   Special devices are often found in the
latter class of plants, whereby diffusion through the stomata
is reduced to a minimum,—the stomata are sunk in pits or
grooves, or surrounded by hairs which tend to maintain a
saturated atmosphere in the neighbourhood of the openings.
A similar reduction of transpiration is attained in plants
which possess the rosette habit, or which grow close to the
ground.

Over and above external conditions which affect the diffusion of water vapour directly, factors causing movements in
the guard cells must be taken into account.   Thus, the stomata
usually close in darkness, so that transpiration virtually ceases
at night.   Reduction of turgor throughout the plant, caused
by failure of a supply of water to the leaves sufficiently rapid
to make good that lost by evaporation, as may occur on a

hot, dry day, also causes closure of the stomata; the rate of transpiration is thereby reduced and the plant is protected from the injury consequent upon excessive loss of water.

Hydathodes—In addition to stomata, many plants possess special water-pores or *hydathodes*, by means of which liquid water escapes from the plant and appears as droplets of liquid upon the leaves or elsewhere. In certain tropical plants the water exuded in this way may be sufficient to cause a steady drip from the leaf tips. The presence of such water-pores can often be demonstrated by keeping plants in a damp, warm atmosphere, thus encouraging absorption and hindering transpiration through the stomata. Plants growing in a warm greenhouse often betray the presence of hydathodes by the drops of liquid which bedew the margins of the leaves. The liquid which exudes may contain soluble substances, and is not necessarily pure water as is that lost during transpiration.

The possession of hydathodes may be of service to plants by enabling them to maintain an upward current of water in the stem, and thus facilitate absorption from the soil when growing under conditions unfavourable to transpiration. They may serve also as " safety-valves," preventing a water-logged condition of the tissues under conditions which favour rapid absorption but hinder transpiration.

DISTRIBUTION OF WATER THROUGHOUT THE PLANT

Water is taken in by the roots of plants and given off or transpired as water vapour from the surface of the aerial parts, especially from the leaves. There is, therefore, a steady stream through the plant, from the roots upwards, known as the *transpiration current*, the rate of which depends upon the rate at which transpiration is taking place. The existence of this upward current in the stem can be shown as follows. Bend the base of a transpiring shoot into a vessel containing water, coloured with a dye such as eosin or red ink. If the bent stem is then severed, the coloured liquid is carried up in the transpiration current, and in a very short time appears in the veins of the leaf, which are stained red by the dye.

We have now to study the process which forms the connecting link between *absorption* and *transpiration*. The problem is twofold: what is the path followed by the water in passing from the root to the leaves, and what forces bring about the upward movement?

Path of the Ascending Water Current—Mention has already been made of the fact that the water absorbed by roots is conveyed to the leaves and other organs through the woody parts of the vascular bundles which form a continuous conducting system throughout the plant. Simple experiments to demonstrate this will be found in the appendix to this chapter. The long tubes formed by the tracheids and vessels of the wood are admirably adapted for the rapid conduction of water. Water travels, also, from one living cell to another by osmosis, but such transference takes place more slowly and is only effective for comparatively short distances.

Terrestrial plants, in which absorption is localised in a root system, and in which, consequently, water must travel from one part of the plant to another, are provided with a well developed conducting or vascular system. Such plants are known as *vascular plants ;* they include the Seed Plants (*Spermaphyta*), also the Ferns, Horsetails, and Club Mosses (*Pteridophyta*). Vascular plants can grow to a large size, and the only non-vascular plants which can compare with them in this respect are certain of the Brown Seaweeds, which live entirely immersed in water. In the latter plants there is no localisation of absorption, since water and mineral salts are absorbed over the whole surface; there is no loss of water vapour from leaves and consequently no necessity for rapid conduction of water over long distances through the tissues. Non-vascular plants which live a terrestrial life are invariably of small size, since the rapidity with which water can be conveyed throughout their living tissues is not sufficient to replace that lost from a large aerial system exposed to dry air. Conversely, in aquatic vascular plants of submerged habit the woody tissues are much reduced, since rapid conduction of water through the plant is unnecessary under such conditions of life.

The Forces concerned in the Ascent of Water—The ascent of water in the stem of a tree or other plant is dependent primarily on forces developed in the leaves as a result of transpiration.

Before discussing the evidence for this statement it is necessary to consider briefly the part played by certain other forces.

Root Pressure—When roots are actively absorbing water from the soil, there may be developed in the absorbing regions a very considerable pressure, described as *root-pressure*. The existence of such a pressure may often be demonstrated, as, for example, when vigorous shoots of a vine or rose are cut off in the spring. Under these circumstances, liquid wells out of the cut surface of the wood in considerable quantities, a phenomenon known as " bleeding." As a result of many observations and experiments, it has been concluded that root pressure, although often surprisingly vigorous, is not sufficiently great to account for the ascent of water to the tops of tall trees. Moreover, measurements of root pressure have shown that its magnitude varies from time to time and shows marked variation in different species of plants; it may be quite small in tall trees such as Conifers, in which water has to be raised to great heights, and is often negligible in amount during summer when the demand for water is greatest.

Capillarity—Capillarity is the physical force which causes a liquid to rise in a tube above the level at which it is standing outside (p. 242). Since wood consists of a system of fine tubes, it has been claimed that this force plays a part in the ascent of water. It can be easily shown by experiment that capillarity will only account for a rise of a few inches in wood and is wholly inadequate to explain the ascent of water in tall trees or the existence of the transpiration current.

Let us now examine the adequacy of the cause suggested to account for the ascent of water, viz. the forces developed during transpiration.

Transpiration—A simple physical experiment will make clear how transpiration may be responsible for the raising of large quantities of water to a considerable height.

Take a thistle-funnel and close the broad end with a layer of plaster of Paris. When dry, fill the tube with water which has been boiled to expel the dissolved air and cooled, and stand it in a dish of mercury with the lower end of the tube dipping beneath the surface of the mercury. Water evaporates from the upper surface of the plaster of Paris, which is kept moist by water drawn from below. Mercury will rise in the tube to take the place of the water lost, and the rate at which water evaporates can be measured by the rate at which the mercury rises in the tube. This experiment can be used to show: (1) that evaporation under the conditions of the experiment causes a " pull " on the water, with the result that the heavy liquid mercury rises for some distance up the tube, (2) that the rate of evaporation, as measured by the rate of rise of the mercury, depends upon external conditions, i.e. it is favoured by dry air, wind, etc. Owing to the force of cohesion between the molecules of water, or what is called the *tensile strength* of water, a very considerable pull can be applied to the column of water without breaking it, so that water or mercury in a tube can be lifted to a great height in this way. With the apparatus described, the height to which the mercury rises is limited either by the formation of a bubble which breaks the continuity of the water column, or by suction of air backwards through the plaster of Paris.

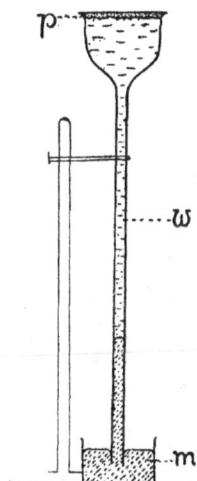

FIG. 9—APPARATUS TO SHOW THE " PULL " EXERTED BY AN EVAPORATING SURFACE OF PLASTER OF PARIS

p, plaster of Paris; *w*, water; *m*, mercury. (*Cf.* fig. 14)

Corresponding to the tube of the thistle-funnel and to the surface of the plaster of Paris, from which evaporation is taking place as described in the experiment, are the vessels and tracheids of the wood in the stem and the living cells of the leaf from the surfaces of which water evaporates into the intercellular spaces. Evaporation from the cells of the leaf causes water to be drawn up the wood of the stem, and the columns of water in the vessels and tracheids of the wood behave similarly to the column of water in the stem of the thistle-funnel (fig. 9).

For this explanation to hold good it is necessary to assume: (*a*) that there are continuous water columns connecting the cells of the leaf from which evaporation is taking place with the cells of the absorbing region of the root; (b) that these water columns do not break under the stress of the " pull "; (c) that the forces developed in the living cells of the leaf are adequate to exert the " pull " required to draw up water to replace that lost by evaporation from the surface of the cells.

The evidence bearing upon these problems is too lengthy to give in full; the conclusions may be summarised briefly as follows.

In regard to (a), the well known fact that air bubbles occur in the vessels and tracheids does not constitute so serious an objection as might appear at first sight. Owing to the structure of the woody elements, air bubbles are kept in such positions that water can find a path round them; they may be regarded as constricting the size of the passage without severing the continuity of the water column.

Experiments dealing with (b) have conclusively shown that a column of water possesses a resistance to rupture, or a " tensile strength," more than sufficient to withstand any strain to which it is likely to be subjected in the plant.

Finally, experiments and observations bearing upon (c) have shown that the osmotic pressure of the leaf cells is sufficiently high to enable them to take in water from the veins of the leaf to replace that lost by evaporation, even though in so doing the pull of gravity on a long column of water has to be overcome. Also, it is known that water evaporates from the surface of living cells under conditions such as those described.

To summarise, the problem of accounting for the ascent of water to the tops of tall trees and the existence of the transpiration current has caused much controversy. The explanations put forward have been broadly of two kinds,—those involving the co-operation of living cells and those resting on purely physical grounds. The latter were based upon experiments which showed that the transpiration current was maintained through stems, the living cells of which had been killed by heat or poisons. The view that has been outlined above, which is that most generally accepted by the plant physiologist to-day, requires the presence of living tissues in the *transpiring* regions of the plant. The living tissues of the root contribute to the process by absorbing water from the soil and thus maintaining a reservoir of water to replace that lost by transpiration. The root-pressure that is sometimes developed as a result of absorption is of secondary importance. Given the physical properties associated with living cells, and a constant loss of water by transpiration, the ascent of water and the existence of the transpiration current can be accounted for on physical grounds alone as a result of the structure of wood and the physical properties of water.

Practical Work

1. Germinate seeds of Oat and Cress in covered dishes on moist blotting paper or in a germinator (see note at end of this chapter). Examine the roots of the young seedlings with a lens and note the root-hairs. (The close relation of the root-hairs with the soil particles can be illustrated by germinating the same seeds in clean sawdust or peat.) Mount the tip of a seedling root of Oat in a drop of water, and examine microscopically. Note the position and extent of the root-hair region and the structure of a single root-hair. Under the high power, note the cell-wall, protoplasm and nucleus. Mount in iodine and note staining reactions. Draw. Mount another root tip in 5 per cent. solution of common salt and note *plasmolysis* of the root-hairs and other cells. Irrigate with water and note whether recovery takes place. Draw conclusions as to the effect of concentrated solutions on roots growing in soil.

2. Investigate the position and extent of the woody tissues in various organs of the plant as follows. Cut across the stem, root, etc. with a sharp knife and moisten the cut surface with a reagent which stains wood—e.g. *anilin chloride* or *phloroglucin*. For this purpose use roots and stems of herbaceous plants and twigs of various trees. (Note: this method does not enable the student to distinguish between *wood* and *skeletal tissue* when the walls of the cells of the latter have become lignified.)

3. Cut shoots of Broad Bean, Maize and young twigs of Lime or other trees. Stand for some hours in a light place with the lower ends in water coloured with eosin. Note the coloration in the veins of the leaves and follow the path taken by the coloured fluid by means of transverse and longitudinal sections through different regions of the stem and leaf stalks. Examine with a lens and make diagrams to illustrate the course of the wood bundles up the stem and into the leaf. In this way can be compared the different arrangement of bundles in the stems of Dicotyledons and Monocotyledons, the different venation of their leaves, and the greater development of wood in trees and shrubs.

4. Take two leafy branches of privet, A and B, and treat as follows. Ring the stem of A, removing all tissues down to the wood; in B, remove the central, woody part of the stem at the base, leaving the outer tissues. Stand the lower end of each in a vessel of water, and place in the light. Compare the rate of

withering in the two branches.   What light does this throw on the course taken by the ascending sap of trees ?   What gardening operations involve recognition of the facts demonstrated ?

5. The existence of root pressure can be demonstrated by taking a vigorous pot plant, cutting off the stem an inch or two above the soil, and attaching a long piece of glass tubing by means of rubber tubing, being careful to make the joints secure by wiring.   Pour a little water into the tube and mark the height at which it stands in the tube; note the subsequent rise of the column of water in the tube due to root pressure.   The pot should be stood on a plate and kept watered during the experiment.

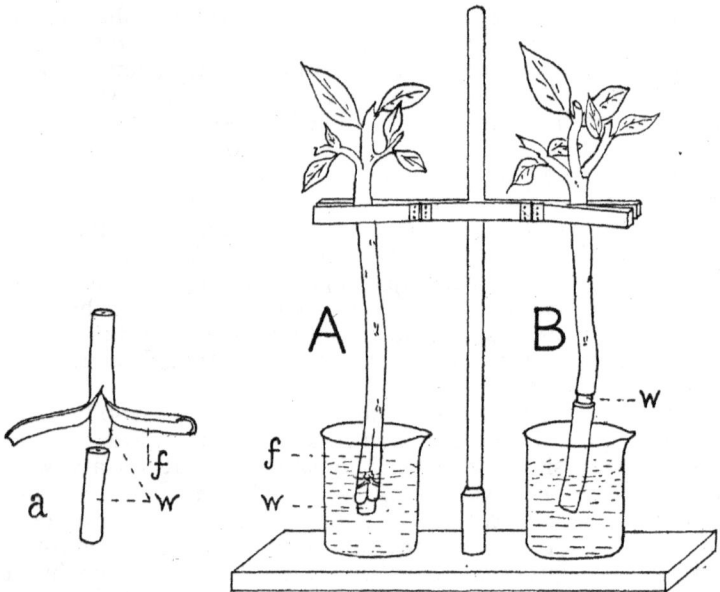

FIG. 10—EXPERIMENT TO DEMONSTRATE POSITION OF WATER-CON-
DUCTING TISSUES IN A WOODY SHOOT

A, woody cylinder and pith removed; B, outer tissues removed, wood
    intact.   w, wood; f, outer tissues of stem.   The smaller diagram, a,
    shows method of preparing shoot for experiment A

6. That there is open communication between the intercellular spaces of a leaf and the outside air can be demonstrated simply as follows.   Select a leaf with large intercellular spaces, as shown by the paler colour of the under side, and smooth petioles—leaves of the common Arum (*Arum maculatum*), Water Plantain (*Alisma plantago*) and Marsh Mallow (*Caltha palustris*) serve well for the purpose.   Attach a piece of thin rubber tubing to the end of the petiole, making an air-tight joint.   Place the experimental leaf in a vessel of water and suck the end of the tubing (or attach it to an air pump) to withdraw air from the intercellular spaces.

As air is withdrawn, water rushes in through the stomata, and the progressive injection of the leaf is easily observed owing to the change in colour.

7. A more elegant method of demonstrating the presence of openings on the leaf surface and the distribution and condition of the stomata is as follows. Make a small glass open bulb like

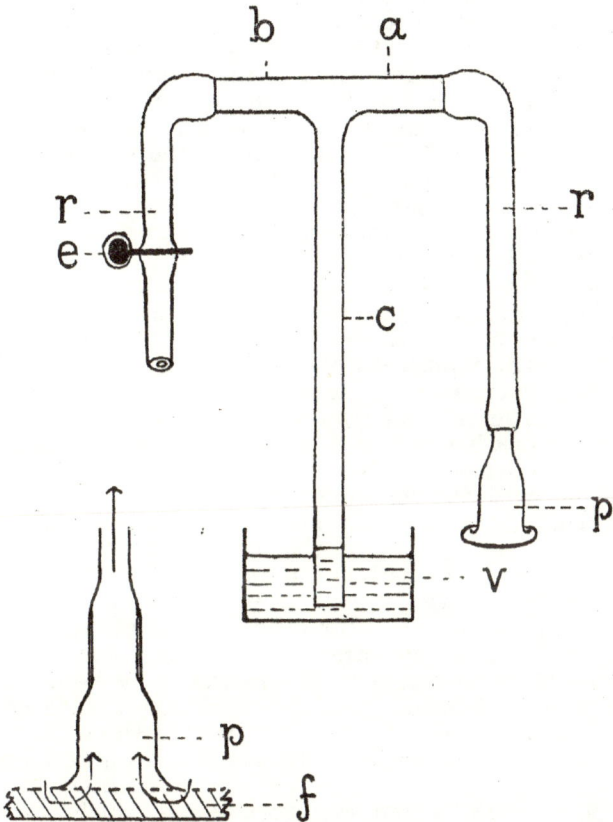

FIG. 11—POROMETER

a, b, c, glass **T**-piece; r, rubber tubing connections; e, clip; p, leaf attachment of porometer; f, foliage leaf; v, vessel containing water. The arrows indicate the course followed by the air when the apparatus is in use

a very small thistle funnel (p), and a **T**-piece of glass tubing as shown in the diagrams. By means of rubber tubing, attach p to one arm, a, of the **T**-piece, and to the other arm, b, attach a piece of rubber tubing r, *making all joints airtight*. Support the **T**-piece in an upright position by means of a retort stand and attach p to the surface of the leaf. In order to make an air-tight joint with the leaf, various adhesives and methods have been tried of which the following is one of the most satisfactory.   Make a stiff jelly with gelatin and allow to cool.   Cut out a thick ring of the gelatin to fit the wide end of p, moisten slightly and attach p to the surface of the leaf by pressing firmly on the gelatin. Glue, also, may be used to make the attachment if care be taken that it is not so hot as to injure the leaf.   To insure success it is important to have the rim of p next the leaf *ground* flat, to choose a portion of the leaf free from project- ing veins, and to be careful not to injure the leaf by heat or pressure.   To use the *porometer*, as this apparatus is called, air is sucked out of the tube r until water has risen some distance up the vertical arm c of the T-piece, whereupon the tube r is clamped by a clip s.   If the area of leaf enclosed within p is without stomata, or if the stomata are closed, the column of water in c remains stationary.   If the stomata are open, air is drawn through them by the weight of the column of water which consequently falls in the tube.   The times taken by the column to fall through a marked distance (x to y) in the tube c are used to indicate the condition of the stomata in the leaves of different plants, and under variations of light, temperature, moisture, etc.   Pot plants can be used for this experiment.   If severed leaves are used it should be noted that the stomata of many plants close temporarily when the leaf is cut or when wilted.   In any case the leaf must be sup- ported in such a position that it is possible to make a perfectly air-tight joint with the porometer (fig. 11).

8. To demonstrate the part played by the stomata in trans- piration, take two leaves of the India Rubber plant (*Ficus elastica*).   Smear with vaseline the upper surface of one leaf and the lower surface of the other leaf and determine the weight of each.   Hang up in the laboratory and, by reweighing at intervals, determine which is losing water more rapidly.   Infer the reason for the difference, and confirm the inference by examination of portions of the epidermis from upper and lower surfaces of the leaves.   Two branches of Laurel bearing a similar number of leaves or any other convenient plant having an unequal number of stomata on the upper and lower surfaces of the leaves may be used in place of *Ficus*, but evergreen leaves which do not lose water too quickly will be found most suitable.

9. Differences in the rate of transpiration under varying con- ditions can be estimated *indirectly* by means of a piece of apparatus called a *potometer*, one pattern of which is shown in the diagram (fig. 12).   A flask is fitted with a two-holed rubber cork, through one hole of which is pushed the cut end of a transpiring branch of Lilac, Laurel, or other convenient plant, and through the other a tube, bent at right angles.   Fill the flask completely with freshly boiled, cooled water, insert the cork and make the joints air-tight by means of wax.   Immediately on filling, place the end of the bent tube under water.   The rates of transpiration can be compared by admitting a small bubble of air to the tube, and observing the time taken for it to pass between two marks

a and *b*.   If this tube is fitted with a stop-cock, and a separating funnel is inserted through a third hole in the cork and filled with water, the bubble can be run back after passing *b* and the apparatus used for a number of observations.   The following precautions should be observed.   Cut the experimental branch under water by bending the part to be severed beneath the surface of water in a vessel.   Stand the branches cut in this way several hours in water before using.   Use boiled water to avoid bubbles and see that all joints are perfectly air-tight.

It is advisable to take the average of a number of readings for each time-record.

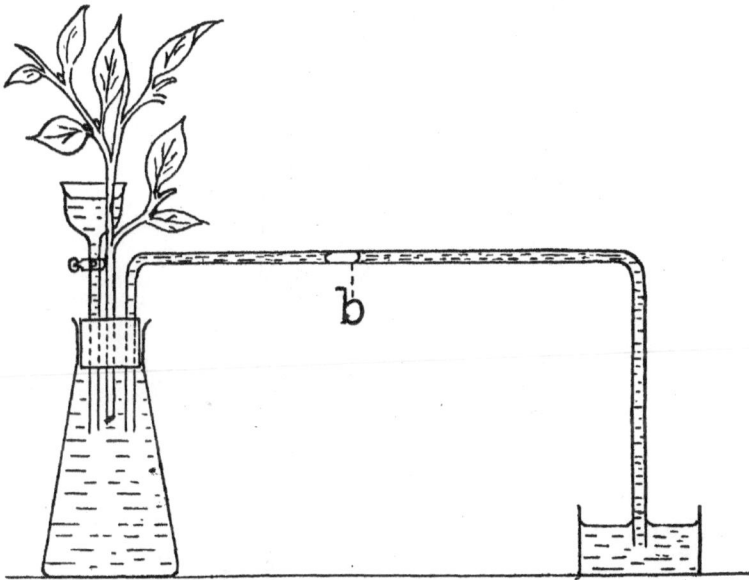

FIG. 12—A CONVENIENT FORM OF POTOMETER

b, bubble

The potometer method is a useful one for comparing the rate of transpiration under varying external conditions, e.g. sunlight, darkness, wind, etc.   It can also be used to show the effect on transpiration of removing a proportion of the leaves, and, by vaselining the upper or lower surfaces of the leaves, to determine from which side of the leaf transpiration is taking place more rapidly.   Another instructive experiment can be performed by determining what proportion of the wood can be sawn through before the rate of transpiration is reduced.

10. The rate of transpiration can be measured *directly* by using a pot-plant, covering the pot and surface of the soil with e.g. oiled silk, weighing at regular intervals, and recording the loss of weight.   To estimate the rate of transpiration for a given leaf

5

area, calculate the area of the leaves as follows. Remove all leaves, lay them on a sheet of paper and cut out a pattern of each leaf in paper.   Weigh a sheet of similar paper of known area and ascertain the weight of the paper leaves.   From these data the *area* of the leaves can readily be calculated.

11. The " pull " exercised by an evaporating surface of plaster of Paris has been referred to in the text (p. 59). The following experiment can be performed in order to compare with this the

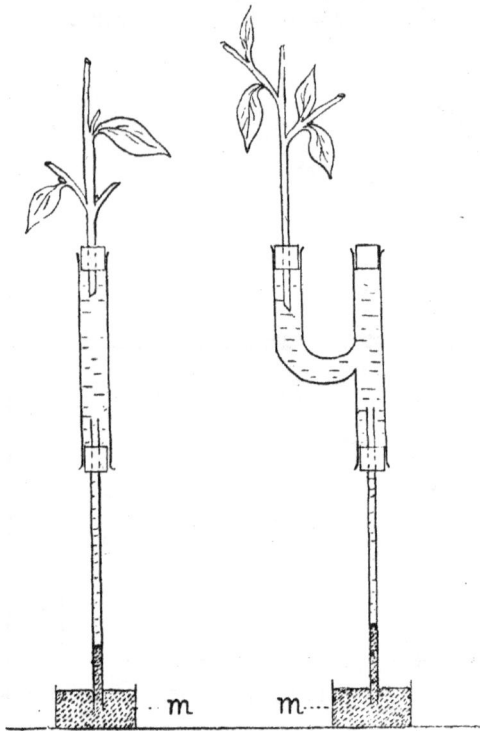

FIG. 13—SIMPLE APPARATUS TO DEMONSTRATE THE " PULL " EXERCISED BY A TRANSPIRING SHOOT

m, mercury. (*Cf.* fig. 9)

" pull " exerted by a transpiring shoot. Attach the lower end of a branch cut as in experiment 9 to a tube of similar bore to that of the thistle funnel (fig. 9), or better, by means of a rubber cork, fix the branch into a tube with a side arm to which the narrower tube is attached by a rubber joint. For direct comparison it is necessary to calculate the area of the leaves as described in experiment 10 and also to measure that of the plaster of Paris. It is instructive to fit up a similar experiment, using a branch from which the leaves have been removed.

12. The existence of a transpiration current can be demonstrated as described in the text (p. 56). Low-growing branches of Lilac or Currant serve the purpose well. The results of this experiment are most striking on a hot, sunny day.

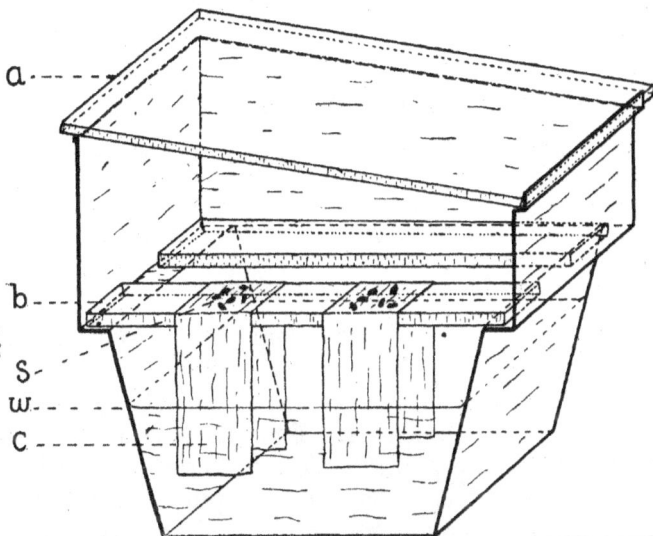

FIG. 14—A CONVENIENT TYPE OF GERMINATOR. (See text)

a, removable glass cover; *b*, glass shelves; *s*, seeds; *w*, water level; *c*, blotting-paper strips

13. Examine the structure of the stomata and their distribution in various plants by mounting small pieces of the leaf epidermis in water and examining microscopically. The epidermis is easily stripped from the following leaves by means of a needle; Broad Bean, Iris, Geranium.

NOTE—A simple and useful form of germinator is shown in fig. 14, and can be constructed of sheet zinc by any tinsmith.

The figure represents a section through the germinator which may be of any convenient dimensions. The seeds are germinated on strips of blotting paper as shown, or in dishes on blotting paper kept moist by other strips of blotting paper dipping into the water. The glass cover can be hinged if a more elaborate apparatus is desired; a sheet of glass as shown serves the purpose. Light can be excluded as desired by a sheet of brown paper.

CHAPTER IV

ABSORPTION OF MINERAL SALTS

CHEMICAL CONSTITUTION OF PLANTS. WATER CULTURES: SAND CULTURES. ESSENTIAL AND UNESSENTIAL ELEMENTS. RÔLE OF MINERAL CONSTITUENTS. THE SOIL SOLUTION; MANURES. SELECTIVE ABSORPTION OF ROOTS: MECHANISM OF ABSORPTION

IN the foregoing pages we have learnt that root-hairs absorb water from the soil, that this water is distributed all over the plant, and that most of it escapes ultimately in the form of water-vapour from the leaves. There is thus a steady stream of water through the plant. We have now to take into consideration the fact that it is not only pure water that plants absorb from the soil, and must inquire what food materials are needed by green plants and whence they are obtained.

There are two methods by which we can discover this; (1) the method of *chemical analysis,* by which we learn exactly what elements are present in the plant, (2) the *experimental method,* by which plants are grown under conditions giving complete control over the nature of the food supplied.

The Chemical Constituents of Plants—Chemical analysis of fresh plant material shows that a large part of the fresh weight consists of water. By drying at 100° C. this water is driven off and the residue can be treated by the chemist in such a manner that he can classify it under two heads,—*combustible material* and *incombustible material* or *ash.* By combustion at a high temperature, the organic materials are decomposed and given off in the form of volatile compounds which can be collected, identified and estimated quantitatively; the small non-volatile residue, representing the mineral substance of the plant known as the ash, can be analysed and its constituents identified and estimated by the methods of inorganic analysis.

The relative proportions of water, combustible material and

ash present in different plants, and in the different parts of plants, is variable as is shown by the figures quoted in Table I.

TABLE I

	Water	Combustible material	Ash
Wheat (grain)	13·65	84·54	1·81
Potato (tuber)	75·48	23·54	0·98
Lettuce (leaves)	94·33	4·64	1·03

As a result of chemical analyses, it has been found that the following elements are always present in plants: *carbon, nitrogen, oxygen, hydrogen, sulphur, phosphorus, potassium, calcium, magnesium, sodium, chlorine,* and traces of *iron ;* with these are usually included a number of other elements. Although such a large number of elements is present, it does not follow that all are necessary for healthy growth. We must distinguish between *essential* and *unessential* elements of the food of plants, and, in order to learn which of those present belong to the former class, we must have recourse to the experimental method.

Water Cultures—Land plants have two sources from which to obtain food, the *soil* and the *air*. The essential feature of the method of investigation known as *water cultures* is that plants are grown with their roots in solutions of known constitution, the plant being entirely dependent upon the solution for supplies of food apart from what it can obtain from the air. There is, therefore, complete control by the investigator over the nature of the food material supplied to the roots. By means of water cultures, it has been shown that many plants can be maintained in health when their roots are supplied with a solution of mineral salts containing the elements *potassium, calcium, magnesium, nitrogen, phosphorus, sulphur,* and a trace of *iron,* in the form of soluble salts. The composition and concentration of a suitable solution has been arrived at by experiment, and it has been learnt that it is necessary, not only to supply certain elements, but to supply them in such proportions and in such concentrations as are acceptable to the plant. It is a remarkable fact that, while solutions of single salts may be poisonous or injurious to plants, the same salts can be combined together so that they can be absorbed without injury by the roots. Such a solution is called a " *physiologically balanced* " solution and those used for water cultures should be of such a nature.

If a plant is grown in a solution from which any one of the essential elements is absent, it sooner or later ceases to grow, becomes unhealthy, and may eventually die if the missing element is not supplied. By means of water cultures it has been learned that the elements mentioned above are essential to the healthy growth of green plants; it has been learnt also that green plants cannot make use of the nitrogen of the air, but that this element must be supplied in the form of an inorganic salt for absorption by the roots. It is to be noted, moreover, that the element *carbon* does not figure among those supplied. With the exception of a few bacteria, green plants are unique among living organisms in that they are able to utilise the carbon dioxide of the air as a source of carbon.

Sand Cultures—It is sometimes convenient to employ a modified form of water culture called a *sand culture*. The experimental plants are grown in quartz sand washed free from all soluble material (for method see appendix) and kept moistened with solutions similar to those used in water cultures. The principle is the same as in the method of water cultures, since there is complete control over the food materials supplied to the roots.

Many plants can be grown more satisfactorily in sand cultures than in water cultures, and the conditions undoubtedly approximate more closely to those existing naturally in soil. The chief drawbacks are that root development cannot be followed so easily as in water cultures, that there is more difficulty in controlling the degree and uniformity of concentration of the solutions in contact with the roots, and that it is more difficult to keep the cultures free from contamination.

Absorption and Utilisation of Mineral Salts—It is easy to show that the elements named are essential for healthy growth, and to note the unhealthy symptoms produced by the omission of any one of them, but it is far more difficult to assign the exact rôle played by each in plant nutrition.

Thus, omission of iron results in failure to form the green pigment chlorophyll, from which it might be inferred that iron is an essential constituent of this substance. This inference would not be correct since pure chlorophyll does not contain iron, and we may conclude that this element is in some way essential to the formation of chlorophyll, although it does not enter into its composition.

A deficiency of certain elements shows itself more rapidly than that of others; conversely, the effect of excessive amounts of salts containing certain elements is very marked. Of such elements none produces a more marked effect than nitrogen: deficiency of suitable salts of nitrogen causes immediate cessation of growth and symptoms of *nitrogen starvation ;* addition of nitrogen, e.g. in the form of nitrates, results in vigorous vegetative growth often at the expense of flower production. For this reason, large doses of nitrates are sometimes given by gardeners to cabbages or other crops grown for the sake of their leaves, but are not supplied, as a general rule, to fruit trees or to " root crops " in which it is important that too much of the energy of the plant should not be expended in the production of leaves. The lowering of general " tone " or vigour, following on a deficiency of certain necessary constituents of the plant food, may be of importance to the farmer or gardener, since it renders plants much less able to resist attack by Fungi or other parasites.

Of the essential elements present in plants, carbon, hydrogen, oxygen and nitrogen are distributed throughout the plant body as constituents of proteins, starch, sugar, cellulose and other organic substances which compose the plant tissues or contribute to their maintenance and growth. Sulphur and phosphorus are found in certain proteins, especially in those of the nucleus. Calcium, potassium, magnesium and iron are distributed throughout the plant, either in the form of inorganic compounds absorbed from the soil, or combined with organic materials. The proportions of these mineral constituents vary with the kind of plant, the part of the plant examined, and the stage of development. Although they constitute a minute proportion of the fresh weight, they are absolutely essential for growth and the maintenance of health. It is possible that minute traces of other elements, not believed to be essential, although constantly present in plants, may also have important indirect effects. *Manganese*, for example, is present in the ash of many plants, and there is some evidence that it has a beneficial effect when supplied in very small quantities. *Silicon*, in the form of *silica*, is of frequent occurrence, and is especially abundant in the ash of Grasses and Horsetails, in which plants it is present in the walls of the epidermis. Although silica is believed to have a beneficial effect when supplied to Grasses or cereals, the evidence is not conclusive that it is an essential element. The same is true for sodium, small quantities of which are almost universally present, and for chlorine, abundant in the ash of maritime plants. The beneficial results which often follow the application of common salt (chloride of sodium) as a manure are probably indirect and depend upon certain chemical changes induced in the soil, as a result of which the supply of potassium available to plants is increased, rather than upon any direct effect of sodium. *Iodine* is extracted on a commercial scale from the ash of Seaweeds although present only in minute quantities in sea-water.

It is known, also, that traces of substances poisonous to living organisms, e.g. compounds of *zinc*, are sometimes present in plant tissues, and that the addition of minute traces of some poisons to experimental cultures may exert a stimulating effect upon growth.

Careful and prolonged researches on animal nutrition have shattered the old view that it is possible to satisfy the needs of development and growth by a supply of carbohydrates, proteins and fats, adequate in amount for the maintenance of growth, repairing of waste, and production of energy. These substances, in a chemically pure condition, do not provide a young animal or a child with the essentials for growth, or suffice to maintain a fully-grown individual in health. There are required, in addition, minute quantities of substances present in small amounts in certain food-stuffs. These have been named "*vitamines*" or "*accessory food factors*," and a great deal is now known about their distribution in articles of food and their behaviour under different conditions, although their chemical nature is still in doubt.

Several accessory food factors are known, distinguished from one another by their distribution in different substances, e.g. fresh fruit and vegetables, eggs, milk, etc., their varying behaviour when heated, and the effects produced upon human beings and animals when they are omitted from the diet. The accessory food factors required by animals are found in various parts of plants and are specially abundant in young seedlings. Their relation to plant nutrition has not yet been fully investigated, but there is evidence that the proper development and healthy growth of plants, like that of animals, is linked up with a supply of the essential vitamines.

Rain, when first formed, consists of pure water, and although rain drops absorb impurities in their passage through the air, the amount of these is so small that the water reaching the ground may be regarded as pure water. Some of the substances contained in soil are soluble in water, more especially when this contains carbon dioxide as does rain water. The film of liquid, with which the roots of plants are in contact in the soil, is not therefore pure water but is a weak solution of mineral salts and other soluble substances.

To the student of plant nutrition it is more important to know the composition of this *soil solution*, with which the roots are in contact, than to know that of the soil itself. Although many methods for the extraction of the soil solution have been tried, no satisfactory way has yet been found by which we can feel assured that the liquid obtained is exactly similar in composition to that of the films of water on the soil particles. Owing to this difficulty, there is still much uncertainty as to the exact composition and concentration of the soil solution. When the chemist makes an analysis of soil he uses, as a matter of convenience, a weak acid for "extracting" the soil and assumes that the solution, so obtained, is similar in composition to that available to plants in the soil. On the other hand, it is known that the elements essential to plant growth are present in all ordinary soils in the form of soluble salts,—chlorides, sulphates, nitrates, etc.,— and that these are constantly renewed by slow weathering of the soil particles.

In the case of natural vegetation, mineral material removed by the roots is returned to the soil by the gradual decay of dead plant material, but in the case of cultivated plants it is removed with the crop, and a shortage of certain mineral substances may result, especially compounds of potassium, phosphates and nitrates. Such deficiency is made good by the addition of manure, either a natural manure, like farmyard manure, by means of which the supply of organic matter in the soil is likewise increased, or artificial manures, which supply directly the constituents lacking. Cultivated soils are especially liable to become deficient in available compounds of nitrogen, owing to the fact that nitrates are extremely soluble and are easily removed in the drainage water, and that there is a constant demand for the nitrates present to supply the needs of plants growing in the soil (p. 93).

It can readily be demonstrated in water cultures that mineral substances are removed from the solution. The roots take up particular substances as required by the plant at the moment and not in relation to their abundance in the solution. This selective action of roots is presumably related to the greater demands made by the plant for some substances rather than for others. The former are constantly removed by osmosis from the root-hair cells in order to supply the needs of nutrition elsewhere, whereas the latter, if absorbed, accumulate and hinder further absorption of the same substance. The whole problem of absorption of mineral salts by the plant is a very complicated and difficult one, and the fact that it takes place cannot be explained merely by reference to the facts of osmosis as can that of absorption of water. The fact that one salt may exercise an influence on the absorption of another is evidence of this (p. 70).

The problem to be explained is not merely that of the passage of mineral substances through the walls of the root-hairs, but is much complicated by the fact that the inner part of the wall consists of living protoplasm, the permeability of which varies from moment to moment; and also to the circumstance that the solutions concerned are in contact not only with the colloids* of the cell but also with those in the soil outside the plant.

The mechanism of absorption by roots is undoubtedly bound up with the physical and chemical nature of living protoplasm through which everything entering the root has to pass. In spite of great advances of knowledge in this direction, our information is not yet sufficiently precise to enable a simple and definite statement to be made, to account

* See Chapter XV.

for all the observed experimental facts regarding absorption of mineral salts by roots.

The root-hair cells are in contact with the living parenchyma cells of the outer tissues of the root. Mineral substances dissolved in the cell-sap of the former can pass from cell to cell by osmosis, until they reach those in contact with the tracheids and vessels of the wood strands, in the central part of the root. It is not easy to account for the passage of water and mineral salts from these living cells to the cavities of the vessels, but it is believed that this is facilitated, at times, by the presence in the latter of osmotic substances such as sugar. On reaching the wood, the water and salts absorbed by the root are carried up by the transpiration current, and reach ultimately the assimilating cells of the leaf. Much of the water evaporates from these cells and is lost from the plant by transpiration; the salts are retained and play their part in the nutritive processes of the plant.

Practical Work

1. **Water Cultures**—Water cultures of small seedlings can be grown in test-tubes or ordinary tumblers. For cultures extending over long periods it is best to use wide-mouthed bottles of not less than one litre (about 1¾ pints) capacity, fitted with clean corks, bored to admit two glass tubes and dipped in melted paraffin wax before use. The following precautions should also be observed.

Clean the bottles thoroughly by washing with strong nitric acid, rinsing repeatedly with tap-water and finally with distilled water. Use pure chemicals and freshly made solutions. Change the solutions every two weeks if possible. Use freshly distilled water which must not be prepared in a copper still; it is desirable to use water distilled on glass.

Into one hole of the cork fix the seedling, keeping the parts above the roots clear of the solution; through the other hole place a short length of glass tubing dipping into the solution below and open above or plugged with cotton-wool to exclude dust. Aerate the solutions twice a week by blowing air through the open tube by means of a rubber bellows, raising the cork meanwhile. Make up any loss from evaporation by adding distilled water. Keep the roots cool and dark by plunging the bottles in sawdust or fibre or by standing them in a box painted black inside, with a closely-fitting lid pierced with holes to admit the necks of the bottles.

Water cultures are best grown in a cool greenhouse or frame. In summer they can be placed out of doors if protected from rain or very hot sun. The following serve well: seedlings of Barley,

Pea, Buckwheat, Sunflower; cuttings of *Fuchsia* or Spiderwort (*Tradescantia zebrina*); plants of the common Duckweed (*Lemna minor*). If seedlings are used, seeds should be selected of uniform size, and germinated on clean blotting-paper. When planting, carefully remove the cotyledons or endosperm and keep the seedlings with their roots in distilled water before placing in the culture solutions. If plants of Duckweed are grown, open dishes or jars must be used instead of bottles, but it is difficult to keep such cultures free from contamination.

Various formulæ are in use for making up the necessary solutions. The following serve well for most plants.

Normal Solution

Potassium nitrate	..	1·0 gm.
Ferrous phosphate	..	0·5 gm.
Calcium sulphate	..	0·25 gm.
Magnesium sulphate	..	0·25 gm.
Distilled water ..	..	1 to 2 litres.

Lacking Nitrates

Potassium chloride	..	1·0 gm.
Ferrous phosphate	..	0·5 gm.
Calcium sulphate	..	0·25 gm.
Magnesium sulphate	..	0·25 gm.
Distilled water ..	..	1 to 2 litres.

Lacking Iron

Potassium nitrate	..	1·0 gm.
Calcium phosphate	..	0·5 gm.
Calcium sulphate	..	0·25 gm.
Magnesium sulphate	..	0·25 gm.
Distilled water ..	..	1 to 2 litres.

Lacking Magnesium

Potassium nitrate	..	1·0 gm.
Ferrous phosphate	..	0·5 gm.
Calcium sulphate	..	0·5 gm.
Distilled water ..	..	1 to 2 litres.

Lacking Calcium

Potassium nitrate	..	1·0 gm.
Ferrous phosphate	..	0·5 gm.
Magnesium sulphate	..	0·5 gm.
Distilled water ..	..	1 to 2 litres.

Lacking Potassium

Calcium nitrate ..	..	1·0 gm.
Ferrous phosphate	..	0·5 gm.
Calcium sulphate	..	0·25 gm.
Magnesium sulphate	..	0·25 gm.
Distilled water ..	..	1 to 2 litres.

Lacking Phosphates

Potassium nitrate	..	1·0 gm.
Ferric chloride ..	..	1 or 2 drops of a watery solution.
Calcium sulphate	..	0·25 gm.
Magnesium sulphate	..	0·25 gm.
Distilled water ..	..	1 to 2 litres.

Lacking Sulphur

Potassium nitrate	..	0·75 gm.
Ferrous phosphate	..	0·5 gm.
Calcium nitrate ..	..	0·25 gm.
Magnesium chloride	..	0·25 gm.
Distilled water ..	..	1 to 2 litres.

2. **Sand Cultures**—Obtain clean quartz sand. Prepare it for use by washing repeatedly with weak sulphuric acid; place under a running tap until all traces of acid are removed and rinse finally with distilled water. Place the sand in clean new flower pots. Select seeds as for water cultures, germinate on moist blotting-paper, plant in the pots of sand, and water with the solutions prepared for water-cultures, giving a measured quantity to each pot at every watering.

N.B.—It must be noted that the effect of withholding compounds of iron from green plants cannot be demonstrated by means of sand cultures, since sand prepared as directed still contains traces of soluble iron salts.

CHAPTER V

CARBON ASSIMILATION

THE NATURE OF CARBON ASSIMILATION, AND THE CONDITIONS UNDER WHICH IT OCCURS: STARCH FORMATION IN THE GREEN LEAF. TRANSLOCATION AND STORAGE. MEASUREMENT OF CARBON ASSIMILATION; "DRY WEIGHT" METHOD. CHEMISTRY OF CARBON ASSIMILATION. EFFECT OF EXTERNAL CONDITIONS ON CARBON ASSIMILATION. ENERGY RELATIONS. LEAF STRUCTURE IN RELATION TO CARBON ASSIMILATION. SUMMARY

Nature of Carbon Assimilation—The plant process of greatest interest to the rest of the organic world is that in which solar energy is absorbed by green plants and stored up in the form of chemical compounds. These storage substances are invariably compounds of carbon and the process, therefore, is one of *carbon assimilation;* since it occurs under the influence of light it is often called *photosynthesis* (φῶs, φωτὸs, light: σύνθεσις, a putting together).

The assimilation of carbon by green plants involves the following operations: (1) carbon dioxide (CO_2) finds its way into the tissues of the leaf from the surrounding air, which contains about ·04 per cent. of this gas: (2) water (H_2O), absorbed from the soil by the roots, is conducted into the leaf: (3) the chlorophyll, or green colouring matter of the leaf absorbs some of the light falling upon it: (4) the energy represented by this absorbed light is used to bring about a chemical action between carbon dioxide and water, with the result that a carbohydrate, sugar, is formed in the leaf, and oxygen is given off as a secondary product of the reaction. A proportion of the sugar so formed is frequently converted into starch *in situ.*

From the above it is evident that carbon assimilation in the living plant can take place only under the following conditions: —(1) when a supply of carbon dioxide is available; (2) when a supply of water is available; (3) in the presence of chlorophyll;

78

(4) when the plant is in the light, since the energy needful is obtained from this source.

In order to confirm these facts experimentally it is necessary to have some method of detecting whether carbon assimilation is, or is not, taking place. This is easily accomplished in the case of plants which form starch in the leaves, for, if we find that certain conditions favour the production of starch in the green leaf, we may presume that they favour also the process of photosynthesis. Use is made of the fact that starch forms a deep blue compound in the presence of iodine. If a leaf containing starch is dipped for a moment into boiling water, soaked in alcohol to remove chlorophyll (since its presence would mask the reaction), and then treated with a weak solution of iodine, the leaf turns a deep blue colour owing to the action of the iodine upon the starch contained in it. The preliminary treatment with boiling water is not absolutely essential, but shortens the time taken for removal of the chlorophyll. Another fact, knowledge of which makes our task easier, is that starch formed in the leaf tends slowly to disappear. If, therefore, a plant, the leaves of which contain starch, is placed in the dark, under which conditions no fresh starch can be formed, that already present in the leaf will disappear: this method gives us a means of obtaining leaves free from starch.

With these preliminaries we are in a position to undertake the experiments described at the end of this chapter. Besides the main facts of photosynthesis outlined above and dealt with in the simple experiments described, there are others of importance which we shall now discuss without attempting to present direct experimental proof.

Translocation and Storage—The starch formed during photosynthesis appears as small grains within the chloroplasts, as may be seen by examining microscopically a thin slice or section cut with a razor from a suitable green leaf or other chlorophyll-containing tissue. When such a leaf is placed in darkness starch gradually disappears, and it becomes pertinent to inquire as to its fate. The answer to such inquiry is that it becomes converted into sugars which dissolve in the cell-sap and are, for the most part, removed from the leaf. The substance responsible for this conversion is the enzyme *diastase* (p. 120). Owing to the presence of diastase in the leaf, there is a continuous conversion of starch into sugar both in the light and in darkness: in the light, however, starch accumulates owing to the fact that its formation during photosynthesis takes place more rapidly than does its conversion into sugar by diastase. Sugar, formed by the green leaf directly or by reconversion from starch, is made use of in three ways: (1) it is used up immediately to supply the needs of local respiration and growth, (2) it is conveyed to other parts of the plant to be used up in a similar way, (3) it is conveyed away to be stored up in various storage organs for future use.

The surplus food manufactured during carbon assimilation is stored in various forms, e.g. as *grape sugar* or *glucose* (Carrot, Turnip), as *cane sugar* (Sugar Beet, Sugar Cane, etc.), as *inulin* (Artichoke, Dandelion and many members of *Compositæ*), and as *oil* (many trees in winter and a number of seeds). The removal of sugar from the leaf to other parts of the plant, whether for immediate or future use, is called *translocation*, a term also applied to any movement of food substances in the plant tissues.

The reconversion of sugar into starch in the storage organs of plants is a result of the activity of special plastids called *leucoplasts* (λευκὸs, white). These plastids resemble chloroplasts except for the fact that they are colourless. The resemblance is more than superficial since it is found that leucoplasts, contained in colourless tissues which grow normally in the dark, turn green when exposed to light owing to the development of chlorophyll. Although in a large number of plants starch is formed both in chloroplasts and in leucoplasts, yet in other plants starch is not found in the leaves, or is present in very small amounts. It may be assumed that the sugar formed by such plants in the light can be used in the leaves or translocated at a rate sufficiently rapid to keep pace with that at which it is formed. The leaves of such plants are unsuited for the experiments described at the end of the chapter. Even in the case of plants in which there is ordinarily an accumulation of starch in the chloroplasts during carbon assimilation, if the rate of assimilation is reduced (e.g. in dull light) below a certain point, starch is removed as quickly as formed and the iodine test applied to the leaves will be negative, although assimilation has actually taken place. It is therefore desirable to have a method for detecting the occurrence and measuring the rate of assimilation, capable of more universal application and greater precision than the iodine test, convenient and simple as is the latter in the case of plants which form abundant starch in the leaves.

Measurement of Assimilation : The " Dry Weight " Method— A suitable method for the purpose is that by which the *increase in dry weight* of leaves during assimilation is measured. The method depends upon the fact that increase in dry weight in a plant is due almost entirely to carbon assimilation. The dry weight of an area of leaf exposed to light is determined. Similar leaves are kept in darkness for a definite number of hours and the dry weight of a corresponding area determined at the end of the period. The value of the latter is less than the former owing to the cessation of carbon assimilation and loss due to respiration and translocation. From the weights obtained the rate of loss per hour of the given area of leaf due to translocation and respiration can be calculated. Call this value " A."   Two similar leaf areas are now selected, one of which is dried and weighed at the beginning of the experiment, and the other treated similarly after exposure to light for several hours. In the latter case there will be an increase in dry weight, due to the fact that carbon assimilation has taken place, and that some of the carbon so " fixed " has been added to the weight. Calculate the amount of this increase per hour and call the value " B." The actual amount of assimilation which has occurred will be greater than " B " since some sugar has been respired or removed by translocation. We have already determined the amount of loss, " A," due to these causes when the plant is kept in the dark. The total amount of carbon assimilation per hour for each area of the leaf will be, therefore, the observed rate, plus the rate of loss from translocation and respiration, viz. " B " plus " A." The processes of respiration and translocation go on continuously in the plant and are unaffected by light or darkness. When proper precautions are taken, this *dry weight method* can be used to measure the rate of carbon assimilation under given conditions with great accuracy and is applicable to all plants and not only to those which form starch in the leaves.

6

Chemistry of Carbon Assimilation—Although the fact of carbon assimilation, characteristic of green plants, is of such fundamental importance to the world in general, the chemistry of the process is still obscure. We know by experiment that under the influence of light, carbon dioxide and water interact in a green leaf to form a sugar and oxygen, and that the amounts of these substances used up and resulting from the reaction are represented by the equation:—

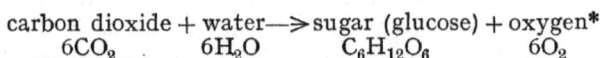

$$\text{carbon dioxide} + \text{water} \longrightarrow \text{sugar (glucose)} + \text{oxygen}*$$
$$6CO_2 \qquad 6H_2O \qquad C_6H_{12}O_6 \qquad 6O_2$$

This equation shows that the volume of oxygen formed is equal to the volume of carbon dioxide used, and that six molecules of carbon dioxide are required for the production of one molecule of sugar. It must not be supposed that the actual chemical change which occurs is so simple as appears from this equation that indicates only the quantitative relations between the participating substances. Sugar is the final product, but it is probably produced as the result of a chain of reactions in the course of which simpler substances are formed; the nature of these intermediate reactions is still in doubt. The reaction, as a whole, requires energy for its accomplishment. Before proceeding it is well to consider what is meant by this statement.

To set a body in motion requires the expenditure of what is called in mechanics *work ;* the greater the mass of the body and the more rapid the motion conferred upon it, the greater must be the amount of work performed. The capacity for doing work is called *energy*, and when this is the result of the motion of a body, as is, for example, the energy of a hammer falling upon a nail, the energy is said to be *kinetic*.† When the energy is not manifest as motion but is " locked up " as it were, ready to appear when conditions allow, it is said to be *potential*. For example, in a catapult, the energy used in stretching the elastic is stored up as potential energy;

---

* This reaction should be compared with that occurring during respiration (p. 30).

† It must be borne in mind that kinetic energy may be manifest, not only as movement of a body as a whole, but also as movement of the individual molecules composing it, i.e. as *heat*, or as movement of the ether, as in *radiant heat, light,* or *electricity*.

when the elastic is released, this potential energy is partly reconverted into the kinetic energy of the shot fired. Potential energy and kinetic energy are thus mutually convertible.

Now chemical compounds possess potential energy, the amount of which varies with the compound. For example, in the reaction by which sugar is oxidised to carbon dioxide and water, the sugar and oxygen possess more potential energy than the resulting carbon dioxide and water. In this change, therefore, some of the potential energy represented by the sugar and oxygen is set free as kinetic energy which may manifest itself in various ways. The reaction between sugar and oxygen to form carbon dioxide and water results in the *liberation* of energy; the reverse change, viz. the combination of carbon dioxide and water to form sugar, requires the *consumption* of energy,—the energy so consumed being stored up as potential energy in the product of the reaction, sugar. Chemical reactions are thus of two kinds: (1) those in which the reacting substances possess potential energy which is set free in some form as a result of the reaction; (2) those which require energy to be supplied before they can take place. The changes occurring in respiration are of the former kind; those which take place in carbon assimilation are of the latter kind.

Chlorophyll—In green plants the energy required for assimilation is obtained from the sunlight absorbed by the chlorophyll. By reason of the part which it plays in carbon assimilation, the nature of chlorophyll is of great interest and has been the subject of much research.

The green pigment which occurs in the chloroplasts and which is spoken of as " chlorophyll " is not a single substance, but is a mixture of several colouring matters or pigments, of which two are green in colour and others are yellow. It is the two green pigments which are responsible for absorbing the light used in photosynthesis.

By means of an instrument called a spectroscope, it can be shown that sunlight consists of many different kinds of light, each of which produces a different colour effect upon the eye, red, orange, yellow, green, blue and violet being those most easily differentiated. These colours are spread out by the spectroscope into a multi-coloured band or *spectrum*. If sunlight or other white light, before reaching the spectroscope, is made to pass through a coloured screen, the particular kind or " colour " of light absorbed by the screen is shown as a gap which appears as a dark band in the spectrum owing to the absence of this colour. The colour of the screen is due to blending of all the remaining " colours " which make up white light and which pass through the screen unabsorbed.

Although chlorophyll is not soluble in water, it can be dissolved out of the green cells by alcohol and certain other solvents, and it is possible by appropriate treatment to separate the green pigments from the yellow pigments with which they are associated.

When a solution of the pure green pigments is examined with a spectroscope, it is found that the green light and yellow light pass through the solution unaltered (giving to it its green colour), while certain kinds of red light and most of the blue light are absorbed. Light is a form of energy, and the light thus absorbed by green leaves represents so much energy captured and available for the chemical *work* of carbon assimilation. It is probable that only a part of this light energy is used in the work of carbon assimilation, and it is not possible to offer any simple or direct explanation of how light energy is converted into chemical energy in the green cells of the leaf. It is known that chemical reactions can be brought about by the action of light as, for instance, when a photographic plate or printing paper is exposed, but it is not possible at present to connect directly in any simple way such *photochemical* reactions with those which take place in the green leaf during photosynthesis.

Since green light does not suffer absorption and so give up its energy in passing through chlorophyll, a plant growing in green light is no better off than a plant in darkness as regards the work of assimilation. In other words, it is only the particular kinds of light which are absorbed by chlorophyll which can be utilised for this purpose.

Since chlorophyll can be extracted from leaves, it might be supposed that by exposure of such an extract to light it would be possible to bring about the combination of carbon dioxide and water (i.e. assimilation of carbon) outside the plant. To test this possibility many experiments have been carried out, none of which have shown conclusively that assimilation can take place outside the plant cell or without the co-operation of the living protoplasm.*

Leaf Structure in Relation to Carbon Assimilation—In studying transpiration, it was learnt that diffusion of water vapour takes place through the stomata of the leaves. Carbon dioxide from the air enters the leaf in a similar manner. In many leaves, e.g. in all those which are exposed to light in a horizontal position, the inner tissues of the upper and lower parts of the leaf are different. Those of the lower part consist of rounded cells bounded by large air-spaces and containing comparatively few chloroplasts; those of the upper region consist of one or more layers of long cells packed closely together like the palings of a fence. The latter or *palisade cells* contain many chloroplasts and compose the chief assimilating tissue of the leaf. The leaves of a Flowering Plant are the principal workshops or laboratories of the plant, and it is in their green cells that the process of carbon assimilation is chiefly carried out. The structure of an ordinary green leaf is illustrated diagrammatically in fig. 15 and it can be judged how well-fitted it is for the functions it performs.

In order that carbon assimilation may take place, carbon dioxide must be brought to the vicinity of the chloroplasts. This is provided for by an extension of the general system of intercellular spaces which is present throughout the plant: in leaves with an upper and lower side, these air-spaces are more conspicuous towards the latter and the colour of the leaf is usually paler on the under side in consequence.

The internal atmosphere of the leaf, surrounding the green cells, is in communication with the external atmosphere through the stomata, through which gases diffuse freely backwards and forwards. From the intercellular spaces, carbon dioxide passes into solution in the water with which the cell-walls are saturated, and thence passes by diffusion into the cells and to the neighbourhood of the chloroplasts. The rate of this diffusion in water depends, as does that of diffusion in air, upon differences in the amounts of carbon

* It is of interest to learn that small quantities of a sugar have been produced in the laboratory under the influence of ultra-violet light, but under conditions so different from those occurring in the plant that no direct assistance towards explaining assimilation in plants can as yet be derived from such experiments.

FIG. 15

A, Structure of a foliage leaf as shown when a thin transverse section is viewed microscopically.

B, Diagrammatic representation of the structure of the leaf, illustrating gaseous exchange between the cells of the leaf and the outside air, in the light and in darkness. Each step of the process is represented by a separate arrow. Thus, in the light, CO_2 reaches the chloroplasts in the following stages : (1) diffusion through the stomata; (2) diffusion through the intercellular spaces; (3) solution of CO_2 and passage through wet cell-walls; (4) diffusion through protoplasm to chloroplasts. The oxygen liberated in photosynthesis escapes from the leaf by similar stages but in the reverse order. Water vapour is given off from the surfaces of all the cells as shown by the dotted arrows.

u.e, upper epidermis; *c*, cuticle; *pl*, palisade parenchyma; *sp*, spongy parenchyma; *s*, stoma; *g.c*, guard cell; *i.c.s*, intercellular space; *ch*, chloroplast; *p*, protoplasm; *w*, cell-wall; *v*, vacuole; *n*, nucleus. The lettering applies to A and B.

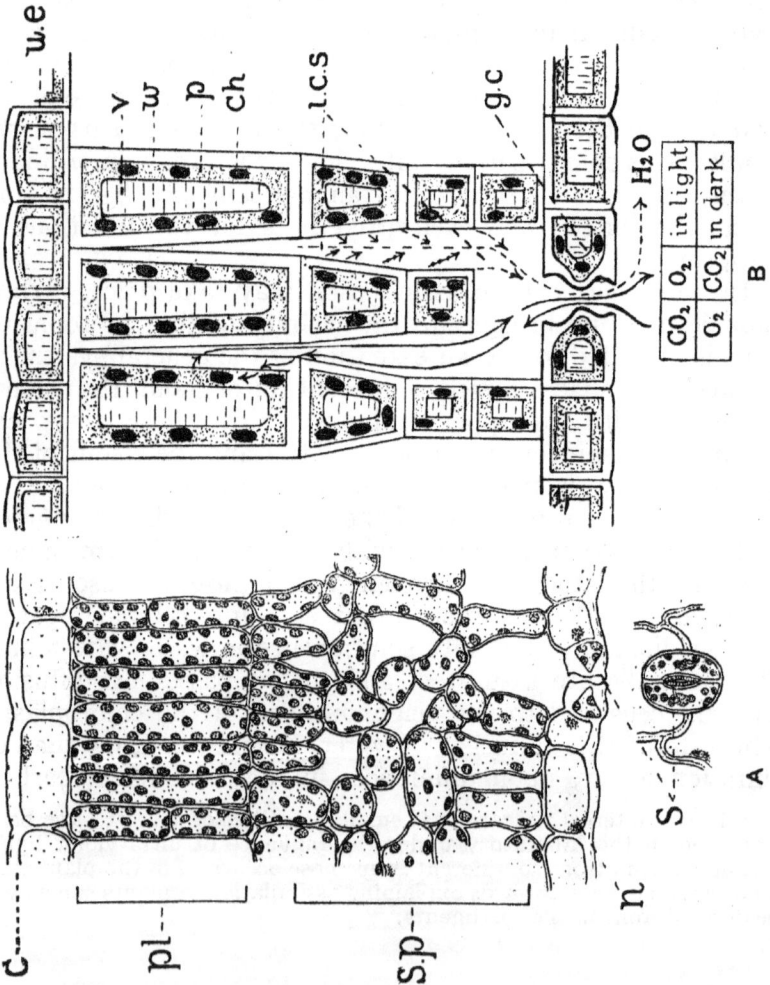

	CO₂	O₂	in light
	O₂	CO₂	in dark

B

A

dioxide present in different parts of the route and therefore, ultimately, upon the rate at which it is used up by the chloroplasts. It must be noted in this connection that diffusion of gases takes place much more slowly in water than in air. On the other hand, movement of the dissolved carbon dioxide, once it is within the cell, may be facilitated by movement of the protoplasm or of the cell-sap.

It must be remembered, also, that the cells of the green leaf, like those of other parts of the plant, are respiring continuously and that the carbon dioxide formed during respiration is available for carbon assimilation during the hours of day-light (see fig. 15).

Consequently, if we desire to measure the amount of carbon dioxide used by a plant during assimilation, it is not sufficient to devise experiments to measure the amount removed from the outside air surrounding the plant, but we must also measure the amount of carbon dioxide given off during respiration in the dark and add together the two values so obtained.

The rate at which carbon dioxide can diffuse inwards depends upon the relative proportions of this gas present in the outside air and in the intercellular spaces, and this in turn is determined by the rate at which carbon dioxide is used up from the latter during assimilation. The rate of assimilation depends upon the species of plant, the condition of the leaves, and external conditions of temperature, light, and the carbon dioxide supply. With the amount of carbon dioxide usually present in the air (about ·04 per cent.), it has been calculated that the leaf of a sunflower plant on a bright day uses up an amount of carbon dioxide equivalent to that contained in a column of air 14 feet high, the area of whose base corresponds to the area of the leaf. The increased weight per day due to carbon assimilation on a fine day in summer is about equal to the weight of a sheet of thin tissue-paper of which the area is equal to that of all the leaves.

Influence of External Conditions upon the Rate of Carbon Assimilation—Under similar external conditions, the rate of assimilation varies in different plants and therefore depends partly upon differences in the leaves themselves.

The effect of external conditions can be estimated by comparing the rates in a suitable plant placed successively under different conditions. Most important of the external conditions which directly affect the rate of assimilation are: (1) temperature, (2) the intensity of the light, (3) amount of carbon dioxide in the air.

We can deal here but very briefly with the results of many researches upon this aspect of the subject.

Up to a certain value, the rate of assimilation in the leaf can be increased by raising the temperature. A point is soon reached, however, when the increase comes to an end, because there is not sufficient carbon dioxide present in the air to satisfy further requirements. In order to carry such an experiment to a conclusion, both the supply of carbon dioxide and the intensity of the light must be artificially increased. Using such precautions, it is found that the rate of assimilation increases continuously as the temperature rises. On the other hand, a high temperature injures the living cells, and such injury is produced more rapidly the higher the temperature. At a high temperature there is, for a time, a rapid rate of assimilation, quickly followed by a fall in the rate as the protoplasm suffers injury. In short, if a leaf is provided with sufficient carbon dioxide, raising it to a high temperature in a bright light leads to a greatly increased rate of assimilation for a time, followed by a fall: the higher the temperature the greater is the initial increase, but the more rapid is the subsequent fall.

Similar experimental results have been obtained in the case of the other factors mentioned.

Thus, given a suitable temperature and sufficient light, the rate of carbon assimilation rises with an increased supply of carbon dioxide up to a certain point beyond which the protoplasm suffers injury. Given a favourable temperature, and an adequate supply of carbon dioxide, the rate of carbon assimilation increases with the amount of light.

It may be concluded, therefore, that in plants growing under ordinary conditions, carbon assimilation takes place most rapidly during bright sunlight on warm days in summer, provided that the plant can obtain sufficient water to satisfy the needs of rapid transpiration; otherwise, the stomata close and the supply of carbon dioxide reaching the inner tissues of the leaf is curtailed.

Increase or decrease in the dry-weight of any plant will evidently depend upon the relative rates at which carbon assimilation and respiration are taking place, since by far the larger proportion of such increase is due to the assimilation of carbon. This being the case, it does not surprise us to find that plants which grow under conditions tending to retard the rate of assimilation, e.g. dense shade, respire at a low rate, since otherwise they would use up food materials more rapidly than these could be replaced and eventually die of starvation.

The function of green plants known as the assimilation of carbon is the most significant fact in the world of organic Nature. Upon it depends directly the food supply of the world and the maintenance of all those vital activities which hinge upon it. Upon it depends indirectly the whole fabric of commerce and industry in our own country and in most modern States. Commercial prosperity in this country is bound up with a supply of coal. Coal represents the accumulated products of carbon assimilation by the green plants of past ages, and coal is the ultimate source of practically all the power used in the machinery of industrial life. A certain amount of power for industrial purposes is obtained from windmills and water wheels or turbines, but in this country at least the amount is inconsiderable. These sources of energy depend indirectly upon the energy radiated from the sun: the green plant is the only means we know at present whereby sunlight can be intercepted and stored up in the form of chemical compounds for the service of man.

Practical Work

1. Cut two leafy twigs of Lilac, stand the cut ends in a vessel of water and leave in the dark for about 24 hours. At the end of this time, test a leaf from each twig for the presence of starch as described in the text. Place one of the shoots in a good light and the other in the dark. (The light in a living-room is not bright enough to give a satisfactory result; the twig should be placed either out-of-doors or in a well lighted greenhouse.) After several hours, again test a leaf from each twig for starch. Plants in small pots may be used instead of cut shoots; small plants of *Tropæolum* (Garden Nasturtium) or Clover serve well for this experiment.

2. In order to show that starch formation takes place locally in a leaf, bind a strip of black paper or a stencil cut in black card loosely on to a leaf (which may be still attached to the plant) so as to shield the covered region from light without excluding air. After exposure to light for a day or so, test for the presence of starch in the leaf in the afternoon. Starch will be found only

in those parts of the leaf which have been exposed to light. By decolorising the leaf and soaking in iodine, a "starch print" of any stencil design may be obtained in this way; a photographic negative having bold contrast may be used in place of the stencil.

3. The necessity for chlorophyll is shown by the following simple experiment.   Selecting a variety of Maple with variegated leaves, take a leaf which has been exposed for some hours to light. Make a sketch of the distribution of the green patches; then kill the leaf, decolorise, and test for starch which will be found only in the regions of the leaf containing chlorophyll.

4. The necessity for a supply of carbon dioxide can be shown as follows.

Take a small leafy shoot of Lilac, or a pot plant, either of which has been kept in the dark until starch-free.   Place the shoot in a vessel of water or the pot plant under a bell jar together with a dish containing caustic soda in order to absorb the carbon dioxide present.   If stood on a sheet of glass, the bell jar can be sealed down with vaseline so that no air from outside can enter. After prolonged exposure to light, the leaves of the shoot do not contain starch, owing to the fact that carbon dioxide has been removed from the air by the caustic soda.   A control experiment should be performed simultaneously with similar apparatus, but omitting the caustic soda and allowing air to enter the bell jar.

5. It is not possible to demonstrate carbon assimilation in land plants by means of the oxygen evolved in the process, since oxygen is a colourless gas.   In the case of water plants, the evolution of oxygen forms a convenient way of recognising that carbon assimilation is taking place, and of measuring its rate.   The method, known as the "bubbling method," can be used as follows.

Take a healthy shoot of some convenient water plant, e.g. the Canadian Pond Weed, tie loosely to a glass rod and stand in a vessel of water so that the shoot is completely immersed, and has its basal end upwards.   On placing in a bright light, bubbles are given off from the cut end, and a stream of bubbles continues so long as the light intensity remains constant.   Increase or decrease in the light intensity leads to increase or decrease in the rate of bubbling, as determined by counting the number of bubbles given off in a minute.   By the use of a photographic exposure meter of the "actinometer" variety, it is possible to demonstrate that the rate of carbon assimilation is proportional to the intensity of the light.

A difficulty is sometimes encountered in this experiment in obtaining bubbles of a convenient and constant size.   This can be overcome as follows.   The cut end of the shoot is dipped in collodion, which, when dry, is pricked with a fine needle, so that an opening is made into the stem of the plant.   This hole remains of constant size, and consequently, the rate of bubbling will be proportional to the rate at which gas is given off.

The "bubbling method" depends upon the fact that water plants take in the gases they require from solution in the water around them. Carbon dioxide is comparatively soluble in water while oxygen is much less soluble in water. Thus, while the carbon dioxide which the plant needs is easily held in solution in the surrounding water, the oxygen set free collects in the tissues and is given off as bubbles.

6. Repeat experiment 5, Chapter II., but germinate the pea seedlings in the light. At the end of a fortnight, dry the seedlings at 100° C. until they cease to lose weight, and compare the dry weight with that of the ungerminated seeds (A), and with that of the seedlings grown in the dark.

CHAPTER VI

THE ASSIMILATION OF NITROGEN BY PLANTS

THE NEED FOR NITROGEN. SOURCES OF NITROGEN AVAILABLE TO GREEN PLANTS: THE NITROGEN OF THE AIR; ORGANIC COMPOUNDS OF NITROGEN; INORGANIC COMPOUNDS OF NITROGEN. THE "NITROGEN CYCLE": THE PROCESSES OF DECAY; NITRIFICATION; NITROGEN FIXATION; DENITRIFICATION. THE ASSIMILATION OF NITRATES BY GREEN PLANTS. SUMMARY

WE have learnt that green plants manufacture carbohydrates from inorganic raw materials, namely, carbon dioxide and water.

A similar state of affairs exists with regard to the nitrogenous materials required for the manufacture of proteins and the building of new protoplasm. We may, in fact, classify plants in two groups according to their mode of nutrition, which is, in the one case, *autotrophic*, in the other case, *heterotrophic*.* In the first group may be placed the vast majority of green plants, which can utilise simple inorganic materials for the formation of the organic substances containing carbon and nitrogen which they require; and in the other group may be placed those plants, destitute of chlorophyll, which depend, as do animals, on supplies of organic materials ready-made.

To the latter belong a majority of the *Bacteria*, all the *Fungi*, and a few species of *Flowering Plants* which have become total *parasites*, e.g. the Broomrape (*Orobanche sp.*) and the Dodder (*Cuscuta sp.*), or total *saprophytes*, e.g. the Bird's Nest Orchid (*Neottia nidus-avis*) and the Yellow Bird's Nest (*Hypopitys Monotropa*). With these plants, which are entirely *heterotrophic* in nutrition, may be included others, which, although green, have adopted to some extent a similar mode of life,—*partial parasites, partial saprophytes*, and that small but remarkable group, the *insectivorous* plants.

* αὐτός, self; τροφή, food; ἕτερος, other.

92

We will now consider the way in which an autotrophic green plant obtains the requisite supplies of nitrogen.

The Need for Nitrogen—Organic substances containing nitrogen are widely distributed throughout the plant, all parts of which contain appreciable amounts of this element. Thus, it has been estimated by chemists that about 2 parts in 100 parts, by weight, of the dry substance of a wheat seed is nitrogen, while that of beans or peas contains about 5 per cent. of this constituent. Leaves and the vegetative parts of plants also contain nitrogen, but in smaller quantity. It can be proved directly that a supply of nitrogen is essential for plants by growing them in water cultures or sand cultures, free from compounds of this element, and observing the rapidity with which growth falls off. In addition, we may learn from such cultures that the free nitrogen of the air is not available to ordinary plants; and also in what forms this element can most acceptably be supplied. The need for nitrogen is also indirectly shown by the rapid growth observable in plants to which it is given in forms such as can be quickly absorbed, and by the poor growth of crops on land known to be deficient in suitable nitrogenous compounds.

The gardener who scatters a little nitrate of soda [$NaNO_3$] or sulphate of ammonia [$(NH_4)_2SO_4$] around his cabbage plants in the early spring expects the vigorous growth by which they respond to this treatment. Indeed, the view may be held that the necessity of obtaining sufficient supplies of nitrogen in a suitable form is a very pressing problem to most plants and to not a few animals. It is not improbable that plants such as *parasites, saprophytes,* and *insectivorous plants* have come into existence, during the course of evolution, as a consequence of the urgency of this " nitrogen problem " as it may be called.

The Sources from which Nitrogen may be Obtained—Three sources of nitrogen are accessible to green plants.

1. The free nitrogen of the atmosphere.
2. Organic compounds of nitrogen.
3. Inorganic compounds of nitrogen.

1. *The Free Nitrogen of the Air*—About 79 parts in 100 parts by volume of ordinary air consists of nitrogen gas.

Nitrogen is a colourless, odourless gas of very inert character, i.e. it does not combine chemically with other elements or compounds at the ordinary temperature, or indeed at all except with difficulty. There are only two ways in which the gaseous nitrogen of the air combines in nature with other substances: by means of electric discharge, causing it to combine with oxygen to form *oxides of nitrogen*, which, as *nitrous* or *nitric acids*, are afterwards washed down in rain; and, as a result of the vital processes of those living organisms which use free nitrogen gas as a food material.

The amount of nitrogen which undergoes change as a result of electric discharge in the air is probably inconsiderable, but it is important to realise that this chemical action may take place, since it has been found possible to combine large quantities of free nitrogen gas on the commercial scale by the use of electricity, so forming compounds which are of great value as plant food when added to the soil.

On the other hand, a number of micro-organisms, especially bacteria living in water and in soil, can use nitrogen gas as a food material. Such organisms are said to be able to " fix " free nitrogen, and their powers in this respect are indirectly of the greatest service to the higher plants.

To the ordinary Flowering Plant, and indeed to a majority of members of the Vegetable Kingdom, the vast supply of free nitrogen of the air is useless as a source of food, a fact which is at once demonstrated by growing them in artificial cultures, certain of which are free from compounds of nitrogen. Plants growing in the latter soon fall behind in growth, show symptoms of starvation and die prematurely from this cause.

2. *Organic Compounds of Nitrogen*—Another source of nitrogen accessible to plants is the *humus* present in the soil and elsewhere. Humus is composed of the decaying remains of plants and animals, and represents great stores of nitrogen and of carbon " locked up " in this way.

Such organic materials are used as a source of nitrogenous food by many of the lower plants, and we know that the successive changes which they undergo in the soil are due to the action of Bacteria, Fungi, and other non-green organisms, which feed upon them and incidentally break them down into simpler substances. To the autotrophic green plant these great stores of organic nitrogen are useless, until they are converted into an *available* form, as a result of the activities of certain micro-organisms.

3. *Inorganic Compounds of Nitrogen*—If we except the small amounts of nitrogen combined as a result of electric discharge, e.g. in thunderstorms, inorganic compounds of nitrogen in Nature are all of *organic* origin, i.e. they are derived from materials which once formed parts of the bodies of plants and animals. The great deposits of nitrate of soda found in Chili, for instance, were formed by the accumulation and subsequent alteration of the excrement of vast numbers of sea-birds.

In ordinary soils, inorganic compounds of nitrogen are present as *compounds of ammonia* and as *nitrates*. Excepting the small amount of nitric acid brought down by rain, these substances are entirely of organic origin, and are products of the decomposition of organic substances in the soil humus.

It has been shown by experiments with water-cultures and in other ways that the green plant is restricted to such compounds of nitrogen as food-materials, and that although certain compounds of ammonia can be taken up and assimilated by some plants, the majority can best obtain nitrogenous food in the form of *nitrates*. From such simple raw materials are built up in the plant workshop the complex chemical substances which we call *proteins*.

The Nitrogen Cycle in Nature—It may be inferred that the manner in which the ordinary green plant has solved the *nitrogen problem* represents one stage in a great cycle of changes which may be spoken of as the " Circulation of Nitrogen " in Nature.

This cycle may be thought of as a chain composed of many links, most of which are vital processes taking place during the nutrition of living organisms. The nature of this cycle may be briefly indicated as follows.

Nitrogen is an indispensable element of the food of animals and plants. There is an inexhaustible supply of free nitrogen gas in the atmosphere. In the gaseous form it cannot be used as a source of food by the higher plants or by animals, but it can be made use of by certain organisms of microscopic size which are widely distributed in the soil and in fresh and salt waters. Such organisms " fix " or combine the free nitrogen of the air by using it as a food from which to build up new organic materials for the growth of their bodies and for reproduction.

In turn, these organisms die or serve as food for larger animals, and in either case organic materials containing nitrogen accumulate in the soil and elsewhere. After death, these are quickly utilised as food by saprophytic animals and plants (p. 247), especially by fungi and bacteria in the soil. Organic compounds of nitrogen are chemically altered, broken down into simpler bodies, and may eventually escape into the air in the form of ammonia or of nitrogen gas, in which case they are temporarily lost to the green plant as food material.

A proportion of the nitrogen locked up in organic substances is continually escaping thus, and it is often possible to smell the ammonia produced during decay.

The Processes of Decay—We speak of the chemical changes undergone by organic substances, after the death of the organisms of which they formed a part, as *decay* or *putrefaction*. We know that these changes are due to living organisms, that they do not happen if these are excluded (p. 251), and that the stages represented are different according to whether the decomposition takes place in the air or under badly-aerated conditions. In the first case *aerobic* and in the second case *anaerobic* organisms are responsible, and the intermediate chemical changes differ accordingly. The part of the nitrogen cycle which most closely concerns the green plant has to do with the stages which intervene between the formation of *ammonia* as a product of the decomposition of organic material, and the appearance of *nitrates* in the soil. The story of the discovery and elucidation of these stages is one of the most fascinating chapters in the history of plant physiology.

Nitrification—Ammonia is a very volatile substance and when formed during the decay of organic material it either escapes at once into the air, whence it may later be washed down in rain, or combines in the soil to form compounds of ammonia, e.g. *carbonate of ammonia*. It is known that some plants can absorb and use certain compounds of ammonia as sources of nitrogenous food. Their importance as food material for the higher plants is, on the whole, insignificant, since, under favourable soil conditions, much of the ammonia formed undergoes changes of a different character.

There is present in fertile soils all over the world a species of bacterium which can use ammonium compounds as a source of nitrogenous food. This bacterium is called *Nitrosomonas* and is a rounded cell of microscopic size which, in the course of its ordinary nutrition, converts *ammonia* (NH_3) into *nitrous acid* (HNO_2), the same substance which is formed in the air when nitrogen is caused to combine with oxygen by electrical means. Nitrous acid combines with any " base " present in soil, e.g. with lime, soda or potash, to form *nitrites*. Also present in most soils is another bacterium, *Nitrobacter*, the small rod-like cells of which have the habit of using *nitrous acid* or *nitrites* as food material, converting these substances into *nitric acid*, which rapidly forms *nitrates* by combining with the bases present in soil.

This remarkable process, taking place in two stages, is called *nitrification*. It results in the capture of much of the ammonia liberated as a product of decay, and its conversion to nitrates which are soluble in the soil water and can be absorbed by the roots of plants. The two species of bacteria concerned in *nitrification* are *aerobes* and the process goes on only when soil is properly aerated.

FIG. 16—NITRIFYING BACTERIA — *Nitrosomonas sp.* (Very highly magnified)

The chemical changes which take place during the conversion of ammonia to nitrous acid and of nitrous acid to nitric acid are *oxidation* changes, and the nitrifying bacteria use these substances in much the same way as ordinary plants use compounds of carbon, i.e. as a means of obtaining energy or as fuel.

It is indeed known from growing artificial cultures in the laboratory that these bacteria, unlike most others, are incapable of using carbon compounds as food and actually object to the presence of such materials where they are growing. Very small quantities of sugar check their activities and for this reason nitrification only goes on actively in soils where the earlier stages of decomposition have taken place, and comes to an end in peaty soils which contain a high proportion of organic matter in the initial stages of decay.

The nitrifying bacteria are autotrophic organisms and resemble the green plant in using the carbon dioxide of the

7

air as a source of food. The carbon so obtained is used for the manufacture of the new plant material required for growth and reproduction. The energy needed for the building up of complex organic compounds from carbon dioxide and water is not, however, obtained from sunlight as in green plants, but is derived from the oxidation or combustion of ammonia compounds: i.e. it is a process of chemo-synthesis instead of photo-synthesis.

Good aeration of the soil by digging, ploughing and other cultural operations is evidently directly and indirectly of the greatest importance to plants.

Directly, because the roots of plants thus obtain adequate supplies of oxygen, because the water supply is improved and the temperature of the soil raised; indirectly, because good aeration is an indispensable condition for the activities of the soil bacteria responsible for the production of nitrates.

Denitrifying Bacteria—Conversely, in badly aerated soil the nitrifying bacteria cease to work and much potentially valuable plant food is lost because of the lack of oxygen.

Members of the soil micro-flora* are not all aerobes. There are present also several kinds of anaerobic bacteria which can use nitrates and ammonia as food in the absence of oxygen. By their activities, these compounds are decomposed and the nitrogen they contain once more escapes in the form of gas. In general, the soil conditions favourable to the growth of the nitrifying organisms hinder the activities of those bacteria which carry on the reverse or *denitrifying* action. The latter are always abundant where organic materials are undergoing decomposition, and their action can best be checked by promoting aeration of the soil by good cultivation.

Nitrogen Fixation—The existence of organisms which use gaseous nitrogen as a food material has already been noted (p. 94) and their importance in the cycle of changes which we are studying must not be overlooked.

It is not improbable that certain species of fungi possess this power, but the process has been most thoroughly studied and the facts accurately determined in the case of certain soil bacteria. These *nitrogen-fixing bacteria* may be classified as

---

* I.e., plants of microscopic size living in soil.

follows: (1) independent organisms living in soil or water; (2) organisms which spend part or the whole of their existence within the tissues of the higher plants.

1. *Nitrogen-fixing Organisms which Live Independently*— Certain bacteria belonging to this class are widely distributed in soil and in water, and one species, named *Clostridium Pasteurianum*, has been studied with special care. It is an anaerobe, often abundant in badly aerated soils, but occurs elsewhere under conditions the nature of which has been carefully worked out in the laboratory. It can be grown in the presence of oxygen if accompanied by two other species of bacteria which are active aerobes and protect it by removing oxygen from the air in its immediate neighbourhood. It is not easy to imagine soil conditions such that oxygen is entirely excluded, but taking the fact just mentioned into account, and also those observations relating to the kind of atmosphere found in soil mentioned in a later chapter (p. 251), we may believe that *Clostridium* can carry on its vital activities even in ordinary well-aerated soil. It is probable, however, that nitrogen fixation takes place under such conditions more rapidly as a result of the activity of aerobic bacteria, especially species of *Azoto-*

FIG. 17—*Clostridium sp.*, SHOWING MANNER OF FORMING SPORES

sp, spores. (Very highly magnified)

bacter, which are common in soil, fresh water and sea water. These bacteria use organic substances such as sugar as a source of carbon, but they are autotrophic with regard to nitrogenous food—i.e. they use gaseous nitrogen.

These nitrogen-fixing organisms are interesting examples of the wonderful versatility shown by Bacteria in respect to nutrition. Bacteria are found practically everywhere in Nature and can exist under widely varying conditions. It is not, therefore, surprising that they show, as a group, much diversity in the nature of their food materials and in the ways they make use of them.

2. *Nitrogen-fixing Organisms which Live in the Tissues of the Higher Plants*—The best known example of this group is the minute organism which inhabits the cells of the swellings or *nodules* on the roots of leguminous plants such as Peas,

Beans, Clovers and Vetches.   In this case the facts are known with certainty: the bacterium has been grown outside the plant, the quantity of nitrogen which it can combine or " fix " in a given time has been exactly measured, and the relations existing between the plant and its invader have been made clear by accurate observations and experiments.

The organism concerned is a small rod-like bacterium, common in most soils, to which has been given the rather cumbrous name *Pseudomonas radicicola*.   These little rods enter the young roots of plants belonging to the Pea family, by forcing their way through the thin wall of a root-hair.

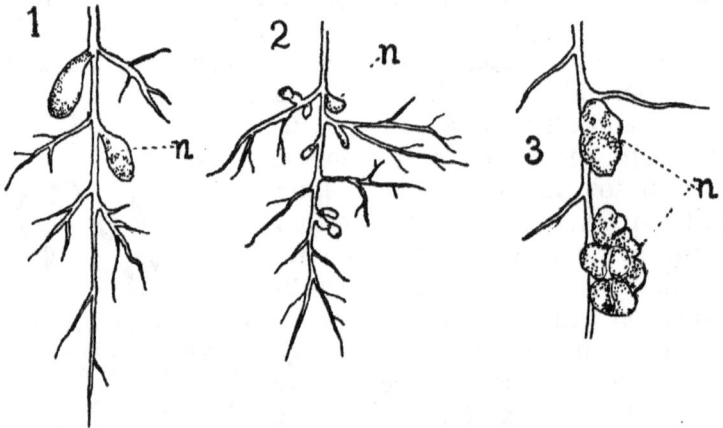

FIG. 18—ROOT NODULES OF VARIOUS LEGUMINOUS PLANTS

1 Gorse (*Ulex europæus*).   2 White Clover (*Trifolium repens*).
3 Broad Bean (*Vicia Faba*).   *n*, nodule

Within the root they multiply rapidly by division and gradually infect the living parenchyma of the root, the cells of which become filled with the bacteria.   The manner of their entry resembles that of many typical parasites which attack the tissues of the higher plants, and the immediate effect upon the plant is similar.   The nodule develops as a result of more active growth of the tissues in the infected region, the living cells of which contain myriads of the bacteria (fig. 18).

The ultimate condition brought about as a result of the infection appears to be undoubtedly of the nature of a " mutual benefit society " with the balance of profit very decidedly on the side of the leguminous plant.   By means of its green leaves the latter makes an abundant supply of sugar, much of which is transferred to the cells of the root. The bacteria profit by the abundance of this ready-made food and use the free nitrogen present in the air of the intercellular spaces of the root as a source of the requisite nitrogenous food material. Thus supplied, they multiply rapidly, but are prevented by the plant from spreading beyond the limits of the tissues of the nodules. Life goes merrily for the invader at this stage; it has found a dwelling place with a rich supply of food, free from the competition of other organisms. At a later stage the tables are turned; many of the bacteria within the cells undergo active digestion by the plant and yield up the nitrogen which has been " fixed " in their tissues. When the plant dies, the roots with their nodules decay, and those bacteria which escape digestion remain in the soil and infect the roots of other leguminous seedlings. The rich stores of organic nitrogen contained in the nodules are ultimately converted to nitrates in the manner previously described (p. 97), and serve to greatly enrich the soil with plant food. This is the practical result from the point of view of the farmer or gardener, and the beneficial effect of growing a crop of leguminous plants was known to practical men long before the details of the mode of life of the nodule-producing organism had been worked out—it was, indeed, known to farmers in the classical times of ancient Rome. It follows that a farmer can successfully grow a crop of Peas, Beans, Clovers, or other plants of the same family on land which is known to be very deficient in nitrates, because these plants can *indirectly* make use of the nitrogen of the air, and at the same time he can feel assured that the land will afterwards contain an increased supply of nitrates for the use of succeeding crops.

Root nodules are not confined to members of the Pea and Bean family. They are found also on the root of the Alder (*Alnus glutinosus*), on the roots of the Sweetgale or Bog Myrtle (*Myrica Gale*), and on those of a number of other plants. In every case in which they have been investigated, the cells of such nodules have been shown to contain either bacteria or the mycelium of a fungus, and it has been inferred from experimental work that the relation between the host plant and its " lodger " is of a similar nature to that described in the leguminous plant.   There seems to be little doubt that many of these curious partnerships between green plants and bacteria or fungi may be regarded as a consequence of the great competition in the soil for nitrates, and the immense advantage obtained by any plant that can draw upon the great stores of nitrogen present in the air.

The exact manner in which compounds of nitrogen formed by bacteria or fungi growing within the cells of a plant are transferred to the latter is not at present quite clear.   In the case of leguminous plants, it is known that the bacterial cells are digested and absorbed, and this also happens in certain cases of that curious association between plant roots and the mycelium of a fungus known as *mycorhiza* (p. 114). There is already a certain amount of evidence that the fungus partner in some such partnerships can " fix " the nitrogen of the air, and when this happens the relation between plant and fungus must be of a very similar character to that between the leguminous plant and the nodule organism.   The matter will be more fully discussed when considering the nutrition of heterotrophic plants in a later chapter (p. 106).

The Assimilation of Nitrates by Green Plants—We know, as yet, comparatively little of the stages intervening between the absorption of nitrates by roots from the soil and their appearance in the form of complex protein substances in other parts of the plant.   It is possible to identify nitrates in many plant organs, and they are usually more abundant in the roots than in the stems or leaves.   It is known that nitrates are not formed in plant tissues unless supplied to the roots, from which it may be inferred that they are not formed by the breakdown of complex substances such as proteins. The materials required for the manufacture of proteins from raw materials include a supply of carbon in the form of either carbon dioxide or carbohydrate, in addition to a supply of nitrates.   Carbohydrates are abundant in the cells of the leaves, and there is a good deal of indirect evidence that green leaves are active centres for the building up or *chemo-synthesis* of nitrogenous substances, as they are for the *photosynthesis* of sugar.

Since proteins are extremely complex substances chemically, it is likely that they are built up in several stages, but there is at present little accurate knowledge as to the details of what happens or as to the nature of the intermediate substances, if any, formed during the process. Moreover, the investigation of such matters is obviously very difficult, because, although it is comparatively simple to identify a substance in a leaf, it is difficult or impossible to ascertain whether such substance is a product of a " building-up," or of a " breaking-down " process taking place there. Nor have we any exact knowledge at present as to the kind of energy which is used in these synthetic processes by which nitrates are converted into proteins.

Two sources of energy are evidently available to green plants; one, the energy absorbed by green leaves from sunlight, the other, the chemical energy liberated during respiration. If the former is used, it might be expected that raw materials such as nitrates would accumulate in those parts of the plant which are without chlorophyll, or are not exposed to light, and there is evidence that this is sometimes the case. There is, however, no evidence at present that the " building-up " process cannot also take place in colourless parts or in the dark and we can only speculate as to the details.

Summary—If we desire to understand the relations of plants to the food supply of animals, it is important to have a clear idea of what is meant by the *circulation of nitrogen* in Nature. This circulation takes the form of a kind of " profit and loss " account, the nature of which is illustrated by the diagram (fig. 19). On the one hand, nitrogen is continually restored to the air by the decay of organic materials, and by the activities of the denitrifying bacteria; on the other hand, it is brought back into circulation by the vital activities of nitrogen-fixing plants, by nitrification in the soil also due to the activities of micro-organisms, and by the green plant which builds up nitrates into organic compounds suited to the need of animals and of man.

Comparing an animal with a green plant in this respect, the former is evidently a " nitrogen spendthrift," whereas the latter is a " nitrogen miser." The animal is extravagant, the plant is economical. The former requires large supplies of ready-made nitrogenous food, because it uses much of this as fuel during respiration, so forming poisonous substances like *urea* and *uric acid* which have to be expelled from the body by means of an elaborate excretory system. The green plant respires non-nitrogenous materials which can be oxidised completely to carbon dioxide and water.

A study of carbon assimilation in green plants teaches us to realise how close is the interdependence between the life-processes of plants and animals. The facts just learnt concerning the assimilation of nitrogen by green plants bring us to the same conclusion. Nitrogen is an indispensable element in the food of animals and of man, as in that of plants; that the former can obtain it in forms which they can use for food depends on the activities of certain micro-organisms, and to a still greater extent on those of the green plant.

Practical Work

(1) **Chemical tests for Proteins**—Make a solution of the water-soluble proteins in pea meal or bean meal by extracting in distilled water, allowing to settle and decanting the liquid.

(a) *Xanthoproteic reaction*—Add strong nitric acid to a little of the above solution in a test-tube; boil, the solution turns yellow. Cool, add ammonia; the yellow colour changes to orange. The latter is the essential part of the reaction.

(b) *Biuret reaction*—To a little of the solution in a test-tube add a small quantity of caustic soda or caustic potash and *one or two drops* of copper sulphate solution. A violet or rose colour is produced in the solution.

(c) *Millon's reagent*—Add a little Millon's reagent to the solution in a test-tube, boil and note the characteristic reddish colour produced.

The above tests may be repeated using proteins of animal origin, e.g. white-of-egg in water, or fresh seeds can be used, ground up and extracted with water or with a dilute solution of magnesium sulphate, since certain proteins insoluble in water can thus be dissolved. The behaviour of such extracts of proteins should also be noted, (a) when heated and, (b) on the addition of alcohol. The green parts of plants are inconvenient for these tests owing to the relatively small quantities of protein present and the masking of colour reactions by chlorophyll, etc.

(2) **Water Cultures**—Grow two sets of water cultures (for method, see p. 75) using Barley, Buckwheat, or other suitable non-leguminous plants. Grow one set in a complete culture solution, and the other set in a solution lacking nitrates, and record the behaviour in each case.

(3) Plant young seedlings of Bean or Clover from which the cotyledons have been removed with a sharp, clean knife in pots of sand prepared as for sand cultures (p. 77), and treat as follows:

Set 1. Water with a complete culture solution.

Set 2. Sterilise the sand and pots by heating in a hot oven for an hour. Water with a culture solution lacking nitrates.

Set 3. Sterilise the sand and pots by heating in a hot oven for an hour; water with a culture solution without nitrates, to the first supply of which has been added a teaspoonful or so of garden soil. Record the rate of growth of the seedlings in the three sets of pots once a week, and at the end of six or eight weeks turn the seedlings out of the pots and examine the roots for nodules.

Set 4. Dig up the roots of as many different species of leguminous plants as possible, wild or cultivated. In the laboratory, wash to remove soil and make drawings to illustrate the distribution and shape of the nodules in each case. The nodules can be sectioned, and sections of the tissues examined microscopically for the details of the cells and bacteria described in the text.

CHAPTER VII

THE NUTRITION OF HETEROTROPHIC PLANTS

THE NUTRITION OF HETEROTROPHIC PLANTS: FUNGI; BACTERIA;
GREEN PLANTS. PARASITES, SAPROPHYTES. INSECTIVOROUS PLANTS.
MYCORHIZA. SYMBIOSIS

A PLANT which is *heterotrophic* in nutrition depends, as does
an animal, upon ready-made compounds of carbon and nitrogen
and is unable to use simple raw materials like carbon dioxide
and water from which to build up organic substances.

These plants are therefore wholly or in part *saprophytes*
(σαπρὸς, rotten; φυτὸν, a plant), using non-living organic
materials as food, or *parasites* (παράσιτος, one who lives at
another's expense), obtaining the requisite food-stuffs directly
from living plants or animals. The plant or animal attacked
by a parasite is called the *host*. The presence of a parasite
often gives rise to symptoms of disease in the host and
eventually may cause its death.

The special properties of the green plant with regard to
nutrition are so bound up with the possession of chlorophyll
that it is not surprising to find that plants which have adopted
a different habit of life in respect to nutrition, cease to form
chlorophyll and often fail to produce green leaves, the organs
by means of which the Higher Plants carry on their characteristic activities.

Fungi — The Fungi are without chlorophyll, and all
members of the group are therefore either *saprophytes* or
parasites, entirely dependent upon the activities of other
organisms for food. There is some evidence that certain
groups of fungi have been evolved from algal plants which
have changed their habit of life, and ceased to produce chlorophyll. Such species show traces of algal descent, not only in
their structure but also in their habit of life, which is aquatic
or semi-aquatic. The Fungi, however, is an immense group

of plants and the vast majority of species are typical land plants, reproducing by means of *spores* adapted for dispersal by air.

Fungi may be of microscopic size, or may produce large and conspicuous *spore-bodies*, such as those of the mushrooms, puff-balls or toadstools. Many of the microscopic forms are *saprophytes*, growing on the remains of animals or plants in all sorts of situations, and such saprophytic species play an important part in the decomposition of organic materials.

Almost any dead branch or twig picked up in a wood in the autumn shows the spore-bodies of fungi and the substance of the wood, examined microscopically, is seen to contain the vegetative threads or *hyphæ* of the fungi responsible for its decay. Saprophytic fungi abound also on organic products such as bread, cheese, or jam, and these materials, if left about in a moist condition, soon form a " seed-bed " on which settle down the spores of such fungi or " moulds " floating in the air. In the case of cheese, the growth of certain moulds is encouraged during the later stages of manufacture, because of special chemical changes they cause in the cheese, and the flavours thereby produced. It is one of the objects of research in scientific cheese-making to learn how to regulate the growth of these species and to control the production of some special flavour desired in the cheese. A vast number of species of fungi are *parasites* which invade the living bodies of other organisms in order to obtain food. Not a few species are parasitic or saprophytic in turn: growing as parasites, they first cause the death of their host, and subsequently continue to live as saprophytes upon the tissues; growing saprophytically upon the dead twig of a plant, they may spread into the living parts and continue growth as true parasites.

Not many parasitic fungi attack animals, although a few diseases of the higher animals and of man are due to their presence, e.g. *ringworm*. On the other hand, a majority of the diseases which affect the Higher Plants are due to attacks of parasitic fungi, the consequences of which are familiar to all as " mildews," " rusts " and other destructive diseases of cultivated plants. Fungi are responsible also for much damage to timber and to timber trees. Parasitic species effect an entry through wounds or damaged places on the bark, grow in and upon the living tissues, and may eventually kill the tree and continue to grow as saprophytes in the dead wood (see Plate I.).

Saprophytic species infect timber after the tree has been felled and, given suitable conditions, especially those resulting from bad ventilation, may cause great damage by setting up " dry-rot " in the wood after it has been used for building.

These are a few examples of the consequences resulting from the *heterotrophic mode of nutrition* shown by the Fungi.

Bacteria—In many respects the Bacteria are the most remarkable group of organisms which exist. Of microscopic size and apparently simple structure, they are found everywhere, and to their activities are due many of the changes continuously taking place in the world around us. Minute organisms of different forms,—small rounded cells or groups of cells, tiny rods or chains of rods (figs. 16, 17), they multiply with the most extraordinary rapidity by repeated division of each cell to form two individuals. Many species also produce spores at intervals, which can withstand adverse conditions and can resist desiccation or extremes of temperature without injury.

Bacteria are by no means typical plants, although some of them resemble the Fungi in respect to their mode of nutrition. Their study is the concern of a special branch of Natural Science, *Bacteriology*, but a brief summary is here included, because, as a consequence of the remarkable variety in their modes of nutrition, they come into touch at many points with the life-processes of plants and of animals.

The cells of Bacteria are usually so extremely small that it is impossible to learn much of their structure even with very high magnification. So far as can be learnt the structure of their cells is very simple; far otherwise is it with regard to the life-processes carried on by these cells, and to this fact is due their great interest to the student of Natural Science. The materials used by Bacteria as food are so varied and the changes wrought in these materials of so remarkable a character, that at every turn our attention is attracted to their activities. The vast majority of Bacteria are heterotrophic in nutrition and are, therefore, like the Fungi, either saprophytes or parasites. Saprophytic species are responsible for the souring of milk, and for many of those changes taking place in food materials left about, which are described as " going bad." These effects are due to the remarkable powers possessed by Bacteria to cause chemical changes in organic materials upon which they are feeding.

Parasitic bacteria give rise to a multitude of diseases. In their search for food, they enter the bodies of human beings or animals,—drawn in with the air breathed, swallowed in the food, or entering the tissues directly through a wound or abrasion of the skin. Within the body, they may confine themselves to some special region or organ, or they may be carried by the blood stream to all parts of the animal. They grow and multiply by division, feeding upon the organic substances around them, either upon the living materials of the cells themselves, thereby causing actual destruction of the tissues, or upon products of cell activity. As a consequence of the chemical changes thus induced, there are formed substances which are poisonous to the host and many of the unpleasant symptoms of disease result, directly or indirectly, from the presence of such poisons or *toxins* within the body, and from the efforts made by the victim to get rid of them.

A number of diseases of cultivated plants are caused in a similar manner. Moreover, the activities of Bacteria do not stop here: although a majority are heterotrophic in nutrition, very many species can use, as food or as sources from which to obtain energy, materials of the most unpromising kind. Thus, although without chlorophyll, the nitrifying bacteria use carbon dioxide as a source of carbon, and obtain energy by the oxidation of ammonium compounds or of nitrites; other species use sulphur or sulphur compounds in a similar manner and are abundant in and about the waters of hot sulphur springs. Some bacteria act similarly upon iron compounds and are responsible for the reddish deposits not infrequently observed near springs or streams; a few can even utilise hydrogen gas as a food material. By learning such facts, we realise how life-processes of the most widely different kinds of organisms are linked together and related to one another, and to the world of inorganic nature in which they take place.

Green Plants—Among the more primitive green plants or Algæ a few species are parasites which attach themselves to other plants, usually to other Algæ, and filch from them such food materials as they require. These parasitic Algæ are usually colourless and, since they are closely related to typical green forms and are exceptional in the groups to which they belong, it is not difficult to believe that the parasitic mode of life is a secondary change and that it has been accompanied by a gradual loss of the power to form chlorophyll. Among Flowering Plants a number of species have become saprophytic or parasitic in their mode of life. According to whether they are entirely dependent upon ready-made food materials or not, they are classified as *total parasites, partial parasites, total saprophytes,* or *partial saprophytes.*

Insectivorous plants are a group of partial parasites of a rather special kind.

Total Parasites—Plants belonging to this group are parasitic upon other Flowering Plants and, according to the organs to which they attach themselves, may be called *root-parasites* or *stem-parasites.* A typical British representative of the former group is the Great Broomrape (*Orobanche major*) a parasite growing upon the roots of Gorse or Broom. A seedling of Broomrape, germinating near the roots of a Gorse or Broom plant, does not develop roots of its own in the soil, but forms rootlike suckers which apply themselves closely to the roots of the host plant, and penetrate the tissues. The cells of the parasite thus come into close contact with those of the host, not only with the living cells of the outer part of the root, but with those of the vascular strands through which food materials are conveyed about the plant.

In this way is obtained ready-made all the food required for growth and reproduction.

The shoot system of an ordinary plant serves the double purpose of producing green leaves, the chief organs of nutrition of the plant, and bearing the flowers and fruits.

In a total parasite such as the Broomrape, green leaves are not formed at all, and the shoot system consists merely of an inflorescence bearing scale-leaves, flowers, and subsequently fruits and seeds. Other species of Broomrape with a similar habit grow upon Ivy, Clover, and other plants. They resemble one another closely in appearance, but differ in their *tastes,* each preferring its own host to that of another species (Plate II).

The Toothwort (*Lathræa Squamaria*) is another parasite of similar habit to the Broomrape, growing upon the roots of

PLATE II

A SPECIES OF BROOMRAPE (*Orobanche hederæ*) GROWING AS A
PARASITE ON THE SOIL ROOTS OF IVY

a, young plant of the parasite ; *b*, mature plant in flower.

The Broomrape is a total parasite without foliage-leaves or soil absorption roots,
and derives all its food-material from its host plant, to the roots of which it is
attached. The species of Broomrape which grows on Ivy is less common than
those found on Gorse and Clover.

Poplar, Willow, Hazel, and other trees. In this plant the root system consists of a number of *suckers* which penetrate the roots of the host and absorb food material from them; the shoot system consists of a branched, horizontally-growing stem or rhizome, bearing curious fleshy scale-leaves, on which arises an erect terminal inflorescence of brownish flowers. As is the case in the Broomrapes, the whole of the special machinery for carbon assimilation is absent: there are no green leaves; chlorophyll is not produced, and the plant relies entirely upon its host for a supply of food. There is a very remarkable family of tropical plants called the *Rafflesiaceæ*, all the members of which are total parasites. In this group, the vegetative part of the plant is reduced to a system of thread-like cells within the tissues of the host, recalling in structure the hyphæ of a fungus. The flowers, often of immense size, appear above ground, emerging from the branches of the host plant in a most puzzling manner.

Of total parasites which attach themselves to the stems of plants, the most remarkable example among British plants is the Dodder (*Cuscuta sp.*), species of which are often found growing upon the shoots of Ling, Gorse and Clover.

The Dodder plant consists of a tangled mass of slender stems, yellowish or reddish in colour, bearing minute, scale-like leaves. The stems of the parasite twine about those of the host and send out suckers which penetrate them. If a thin section is cut through the junction, the tissues of the two plants are seen to be in intimate contact: union with cells of the wood provides for the transference of water and salts, union between the tissues of the bast provides for the absorption of ready-made organic materials. In late summer the tangled stems of the Dodder bear dense clusters of small pinkish-white flowers, and usually produce an abundant crop of seed.

It is interesting to compare such total parasites with the ordinary Flowering Plants to which they are most nearly related. The flowers of Dodder closely resemble those of Convolvulus and both plants have a twining habit of growth. There are indeed so many points of resemblance that the Convolvulus and the Dodder are placed in the same family *Convolvulaceæ*.

The peculiarities of appearance and structure in the Dodder

are evidently related to its special habit or mode of nutrition
and it is not difficult to find evidence that they are of a secondary kind.   Thus, leaves are produced, but they are very
small and have evidently ceased to function in the nutrition
of the plant.   Traces of chlorophyll can be found and the
structure of the flowers, fruits and seeds agrees with that of
more typical members of the *Convolvulaceæ.*   It is likely that
the ancestors of the Dodder were twining plants with green
leaves, but with adoption of the parasitic mode of life, they
ceased to produce the machinery necessary to the existence
of an independent green plant.   Not only are green leaves
absent, but the Dodder plant has lost the power of forming
soil roots.   It is very interesting to germinate seeds of the
parasite and watch the behaviour of the seedlings if unable to
attach themselves to a suitable host-plant soon after they
germinate.

Partial Parasites—Another proof of the secondary nature
of the parasitic habit in Flowering Plants is provided by the
existence of a number of plants belonging to different families
which are *partial parasites.*

The Mistletoe (*Viscum album*) is a case of this kind.   Perched
upon the branches of a tall tree of Elm or Poplar, it is entirely
cut off from the soil and must obtain the necessary water and
salts from its host, to which it is attached by an elaborate
system of " suckers " which penetrate the wood of the
branch on which it is growing.   The Mistletoe, although
rootless, has green leaves and it may be suspected, although
there is no direct evidence, that this plant represents a stage
on the way to complete loss of independence such as is shown
by the Dodder.

There are also numerous cases of plants which are *partial
root-parasites* owing to their habit of forming special suckers
which penetrate the fine roots of grasses and other plants and
filch food materials from them.   Among such are many
members of the Foxglove family; e.g. the Yellow Rattle
(*Rhinanthus Crista-galli*), the purple Bartsia (*Bartsia Odontites*),
the Eyebright (*Euphrasia officinale*), and a number of others.

When studied experimentally, plants of this kind show
different degrees of dependence upon their hosts, and thus
mark progressive stages of adaptation to the parasitic habit
of life.

Insectivorous Plants—Such common insectivorous plants as the Sundews (*Drosera spp.*) and the Butter-Wort (*Pinguicula vulgaris*) are known to everyone. As regards their nutrition, these plants are partial parasites. They possess green leaves and can carry on carbon assimilation, but it is significant that their root-systems are scanty and badly developed and that they grow naturally in wet, peaty soils likely to be deficient in nitrates. These plants catch small insects which alight upon them, by means of sticky hairs upon the upper sides of the leaves. The cells of the hairs form and excrete special protein-digesting enzymes (p. 121), very similar to those present in the stomach of an animal during digestion. By means of these enzymes the flesh of the trapped insect is rendered soluble, and the products of digestion are subsequently absorbed by the hairs of the leaf. Among other plants with the insectivorous habit are the Pitcher Plants, the leaves or part of the leaves of which form pitcher-like structures. These pitchers serve as death-traps to insects, which fall in and are drowned in the liquid which collects at the bottom of the pitcher. In the case of some of these plants e.g. species of *Nepenthes*, there is evidence that the plant digests the bodies of its victims by means of special enzymes. In other cases, e.g. in the American Pitcher plants (*Sarracenia spp.*) and the Bladder-Wort (*Utricularia spp.*), there is no evidence that this is the case, but the plant probably profits by absorbing the soluble products of decay due to the action of various micro-organisms present in the liquid of the pitchers.

Insectivorous plants are evidently heterotrophic in respect to their nitrogenous food, which is obtained and dealt with by some of them in exactly the same manner as is that of a carnivorous or flesh-eating animal.

Total Saprophytes—There are a few Flowering Plants which are apparently entirely dependent upon dead organic material for the carbon of their food. British plants belonging to this class are the Bird's-nest Orchid (*Neottia nidus-avis*), the Coral Root (*Corallorhiza innata*), and the Yellow Bird's-nest (*Hypopitys Monotropa*).

These plants have much the same habit as a total root parasite such as the Broomrape as regards their aerial parts, but they possess an elaborate underground system formed by the repeated branching of roots, or,—in the case of the Coral-root Orchid,—of an underground rhizome.

These plants grow beneath trees in soil consisting almost entirely of leaves in various stages of decay. They are therefore in contact with an abundant supply of organic material containing carbon and nitrogen. On the other hand, leaves are represented only by small, brownish scales, they form no chlorophyll, and are likely to lack nitrates owing to the situations in which they grow. The nutrition of these plants is evidently heterotrophic, but it is not easy to discover its exact nature and we know curiously little about the details.

They are all *mycorhiza-plants*, and the tissues of the underground parts, whether roots or rhizomes, are invariably invaded by the hyphæ of a fungus (p. 102). The *mycorhiza* has been very carefully examined in the plants mentioned, by means of sections of their roots or rhizomes, and the cells show certain remarkable features. In some cells, the hyphæ of the fungus are evidently growing actively, whereas in others they form an apparently structureless mass, and the two conditions can be related to one another by cells showing an intermediate type of structure. These appearances are interpreted as follows. The fungus invades the root cells at a very early stage of development and grows at the expense of the food materials contained in them. Sooner or later, certain cells of the root *digest* the hyphæ which thus yield up the stores of food which they contain. The hyphæ within the plant cells are directly connected with others growing in the soil outside the roots. It is probable that the fungus can utilise organic compounds present in the soil as a source of food, and that by digestion of the hyphæ within the root cells, the plant indirectly derives benefit. It is not, however, at all clear in what way the fungus profits by the relationship !*

Partial Saprophytes—A large number of plants form mycorhiza and in certain families, notably in the Orchids and in the Heath family, the relationship between the plant and its fungal partner has been carefully worked out (p. 115).

There is little doubt that in many, if not all these cases an exchange of food material goes on between plant and

---

* A remarkable case has recently been described in which a Flowering Plant, the Orchid, *Gastrodia elata*, is parasitic upon a Fungus (*Armilleria mellea*) for a part of its life-history.

fungus, and that such mycorhiza-plants may be partial saprophytes drawing indirectly upon the organic materials in the soil by means of their fungus-partners. The same is evidently true of plants which obtain organic food from the organisms living within the tissues of their nodules. To give all the evidence for this view demands more space than is at our disposal, although the facts are of great interest in a study of plant nutrition. In the majority of Orchids possessing green leaves, the relations of the plant with its root fungus, as interpreted by sections of the roots, are similar to those described for the Bird's-nest Orchid. In certain cells of the underground parts the fungus grows vigorously, and in these green species can presumably profit by a supply of carbo-hydrates from the leaves; in other cells of the root, the fungus undergoes digestion, and thus yields up the food materials contained in the hyphæ for the use of the plant.

There is no evidence that the fungi which grow within the roots of Orchid plants can utilise the nitrogen of the air, but they presumably can draw upon the organic materials of the humus in the soil around the plant for this constituent of their food.

The most interesting feature of the mycorhiza of Orchids is the fact that its formation has become indispensable to the plants. Orchid growers have always found difficulty in germinating seeds successfully, especially those of certain species. It is now known that in a great majority of cases the seedling cannot develop beyond a certain point, unless its tissues are invaded by the fungus-partner at a critical stage which varies with the species, i.e. the plant owes its very existence to the presence of the fungus.

This discovery explains why Orchid growers found it useful to mix pieces of the roots of the parent-plant with the soil in which they sowed seeds. The fungus, present in the former, could spread through the soil and invade the young seedlings as they began to germinate.

In the Heath family, a similar state of affairs has been shown to exist in the common Ling (*Calluna vulgaris*). In this case also, the seedling cannot grow into a plant unless the fungus present in the roots invades the tissues soon after germina-tion; i.e. the partnership has become indispensable to the green plant.

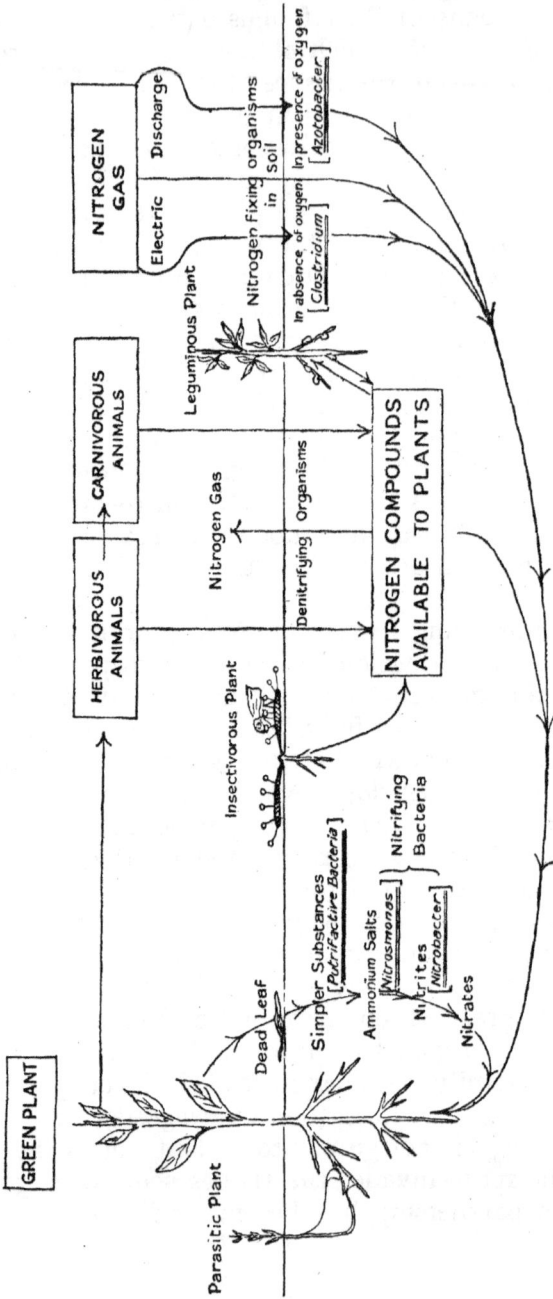

FIG. 19—DIAGRAM ILLUSTRATING THE "CIRCULATION OF NITROGEN" IN NATURE

The plant in this case does not risk the chance of non-infection by the appropriate fungus from the soil, for the seeds carry the hyphæ of the fungus-partners upon the seed-coat and the fungus becomes active as soon as germination takes place.

Symbiosis (Greek σύν with, βίος life)—This term has been used by botanists and others to describe such relationships between living organisms as are apparently of the nature of " mutual benefit societies." Of such a character are the curious partnerships between Flowering Plants and Fungi, just described. They are quite comparable with the conditions found in the *Lichens*, in which each so-called species is really a dual organism, consisting of a fungus and a green alga living together in close association. The green plant assimilates carbon from the carbon dioxide of the air, and is sheltered from desiccation by the more resistant tissues of the fungus; the latter obtains carbon compounds from the alga.

The term *symbiosis* means " living together," and is used to describe cases in which there is presumably an interchange of food materials with some degree of mutual benefit. As knowledge of symbiosis among plants is extended, it becomes clear that the conditions existing between the partners vary greatly, not only in one case as compared with another, but from time to time in any one case. Just as in human society we find partnerships of all kinds—those in which one partner profits greatly at the expense of the other, as well as those in which the relation is constantly one of mutual benefit—so in these *plant partnerships* many degrees of co-operation are represented, and there is much variation in the degree of stability of the relationship.

Practical Work

1. Dig up a few of the superficial roots of Beech or Scotch Fir. Wash carefully and select roots with uninjured tips for examination. Cut off about ⅛ of an inch from the tip, mount in a drop of water and examine microscopically. Note the dense web of fungal mycelium closely investing the root (*ectotrophic* mycorhiza).

2. The presence of fungal hyphæ within the tissues of roots (*endotrophic* mycorhiza) can be well seen in sections through the soil roots of any Orchid.

CHAPTER VIII

ENZYMES

THE reader may have noticed that an important point concerning the chemical changes which occur in plants has been left unexplained. In dealing with respiration we have spoken of the " breaking-down " of sugar with the production of carbon dioxide and alcohol, or of its oxidation to carbon dioxide and water. In discussing the assimilation of carbon and the germination of starchy seeds, it was said that starch is converted into sugar. Now, at ordinary temperatures, these changes do not take place as the result of simple chemical action. How, then, are they brought about in the plant cell ? The answer to this question is that they result from the action of *ferments* or *enzymes*. It becomes necessary, therefore, to inquire into the nature of these bodies known as enzymes and to ascertain the conditions under which they can manifest their activities. We have learnt (p. 32) that living Yeast can ferment a solution of sugar, with the production of carbon dioxide and alcohol, the Yeast plant using the energy liberated by this reaction for its life processes. If the Yeast cells are killed by treatment with strong alcohol or ether, and the dead cells filtered off from the liquid and added to the sugar solution, the same chemical changes will occur, although the Yeast, being dead, does not grow and increase in quantity. The fermentation of sugar (glucose), with the production of alcohol and carbon dioxide, does not therefore require the presence of living protoplasm for its accomplishment and is brought about by some non-living substance or substances contained in the Yeast cell. This non-living substance is an enzyme called *zymase*.

It is characteristic of enzymes that a small quantity can bring about a relatively great amount of chemical change if given sufficient time, and also that the enzyme is but little affected or used up in the reaction. In these respects enzymes resemble the inorganic bodies known to chemists as *catalytic agents*, or *catalysts*, in so far as they appear to accelerate a reaction without themselves being destroyed.

A catalyst can only initiate or hasten a reaction which does not require the consumption of energy (p. 83). Its action has been compared with that of oil upon a stiff machine,—the oil in such a case allows the machinery to move, or increases its rate of movement, but does not provide any energy for the purpose. It is to be noted that oiling a machine frees its movements in both directions, whether it goes backwards or forwards depending not on the oil, but on the forces driving the machine. Similarly, an enzyme that can hasten a reaction in one direction can almost certainly hasten it equally in the reverse direction, if the conditions are favourable to the reaction proceeding in that way.

We do not know the exact chemical constitution of any enzyme, since it has not been found possible to obtain these bodies in a pure state, but, judging from their behaviour under certain conditions, it is believed that chemically they resemble proteins. Thus, they lose their characteristic properties by boiling, as do many protein solutions, and they can be thrown out of action by certain poisons which destroy living protoplasm. On the other hand, enzymes can be preserved in a dry condition for a long time, and can be precipitated from watery solutions by the addition of alcohol or ether without losing their properties. Consequently, Yeast, killed with alcohol, after it has been boiled loses its property of fermenting sugar, but if dried, may be stored and used at any subsequent time.

Another enzyme found in Yeast is that called *invertase* which causes the splitting or " inversion " of cane sugar into two other sugars,—glucose and levulose. The change which takes place is expressed in the chemical equation:—

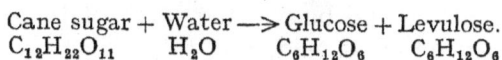

$$\text{Cane sugar} + \text{Water} \longrightarrow \text{Glucose} + \text{Levulose.}$$
$$C_{12}H_{22}O_{11} \qquad H_2O \qquad C_6H_{12}O_6 \qquad C_6H_{12}O_6$$

This change evidently involves the using up of water and is described in consequence as *hydrolysis*. Many of the chemical changes wrought by enzymes are of this nature. In the present instance, cane sugar gives rise during the reaction to two different sugars of simpler chemical composition. These two sugars, glucose or dextrose and fructose or levulose, are not identical in chemical constitution and properties, although they contain the same chemical elements in similar proportions. The enzyme invertase differs from most plant enzymes in that it can pass readily through cell-walls and so escape from the cells after their death. If some Yeast, which has been killed by soaking in alcohol, is shaken up in water which is then decanted off, the water will possess the property of inverting cane sugar, although not of fermenting the resulting glucose: the latter property remains with the Yeast cells which still contain the enzyme zymase (p. 118). Zymase may be obtained in solution, however, by grinding Yeast with sand, which ruptures the cell-walls and allows the enzyme to escape. It is clear, therefore, that the capacity of Yeast to ferment cane sugar depends upon the fact that each Yeast cell produces two enzymes which act upon sugars; one of these, *invertase*, changes cane sugar into glucose and levulose, the other, *zymase*, breaks down glucose to carbon dioxide and alcohol. Zymase is unable to act directly upon either cane sugar or levulose.

The inversion of cane sugar into glucose and levulose can be brought about outside the plant by chemical means, e.g. by boiling a solution of cane sugar with a small quantity of strong mineral acid. The acid in this case acts as a catalyst, but its action is extremely slow compared with that of invertase and only occurs at a measurable rate at high temperatures.

Another important enzyme, widely distributed in plants, is *diastase*, which converts starch into sugar. This reaction, like that brought about by invertase, is one of hydrolysis, and can be expressed thus:

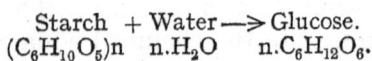

$$\text{Starch} + \text{Water} \longrightarrow \text{Glucose}.$$
$$(C_6H_{10}O_5)n \quad n.H_2O \quad n.C_6H_{12}O_6.$$

The same change can be effected in the laboratory by the use of a strong mineral acid, but takes place with extreme slowness at ordinary temperatures.

In the plant, the conversion of starch into sugar is always brought about by the action of diastase. One of the occasions when this becomes necessary is during the germination of starchy seeds. The seed at germination undergoes very active growth which necessitates rapid respiration; a supply of sugar is required and is derived from the starch stored up in the seed. Starch cannot be translocated, since it is a solid; it is therefore converted into sugar by diastase and transferred in solution to the growing regions (p. 35). Diastase is also responsible for conversion into sugar of starch formed in green leaves as a result of carbon assimilation (p. 79).

In reactions brought about by enzymes, as in other chemical changes, accumulation of the products of the reaction tends to decrease the rate of the reaction, or may bring it to an end. Thus, the rate at which diastase converts starch into sugar depends upon whether the sugar formed is removed or accumulates. In a green leaf, the extent to which starch accumulates during carbon assimilation depends upon the rate at which it is converted into sugar, which, in turn, depends upon the rate at which the sugar is translocated away from the assimilating cells.

Other enzymes of great importance found in plants are the protein-splitting enzymes (*proteases*), the fat-splitting enzymes (*lipases*) and the oxidising enzymes (*oxidases*).

Insoluble protein reserves occur in seeds and other parts of plants; converted into soluble substances by protein-splitting enzymes they can be then carried in solution to those parts of the plant where they are needed. The " digestion " of protein material by an insectivorous plant and by animals is brought about in a similar manner. The fat-splitting enzymes play a similar part in rendering soluble the oils and fats stored in many seeds and reserve tissues. The appearance of a brown coloration upon cutting or injuring certain plant tissues such, for example, as the fruit of many varieties of Apple, is due to the presence of oxidising enzymes. These tissues contain colourless substances, called " chromogens," which become oxidised to brown substances under the action of oxidising enzymes or *oxidases*, when the cells containing them are exposed to the air. The colours of many flowers also are due to the action of oxidases upon colourless chromo-gens, leading to the production of pigments of various colours, according to the nature of the particular chromogen con-cerned.

Sometimes an enzyme and the substance upon which it acts are present in different but neighbouring cells of the same tissue. In such cases, mechanical injury of the tissues or death of the cells may allow the two bodies to come together and interact. Bruising the seeds of bitter almond, for example, allows the enzyme *emulsin* to come into contact with the substance *amygdalin* occurring in other cells of the seed, with the result that the latter is decomposed into several simpler substances, one of which, prussic acid (hydrocyanic acid), can be readily recognised by its characteristic smell.

Without particularising in greater detail, one may sum up by the general statement that the majority of the chemical reactions taking place in the cells of living organisms are brought about by the co-operation of enzymes. Consequently, since enzymes are much more specific in their action than chemical catalysts, one enzyme usually acting on only one or a few substances, the number of different enzymes found in plants at one time or another is very great. The manner in which the production of these enzymes at the appropriate times is regulated and their activities controlled is at present imperfectly understood. Although, for the most part, the activities of protoplasm can be conceived as the result of a number of fairly simple chemical reactions initiated by various enzymes, the mechanism controlling the production and activities of these enzymes still remains one of the mysteries associated with life.

General Summary—We began our study of plant nutrition by comparing the plant with a machine like an engine, and the simile proved useful, although it cannot be pushed to its extreme limits. A plant, like a machine, depends on the outside world for materials required for three purposes: (1) for initial construction, or *growth*, (2) for *repairs* during active life, and (3) for a supply of *energy* by which it can do *work*. In the case of an organism, the work done is made manifest in the activities of *life, growth* and *reproduction*, and the general mechanism by which the energy for these is provided is called *nutrition*. As in the machine, so in the plant, all the materials needed for the manufacture of the one and the development of the other are not necessarily required afterwards. It is known, for example, that small amounts of certain substances are indispensable to the proper development of a young animal and can be dispensed with at maturity, and the same is probably true for plants.

The changes involved in plant nutrition are of two kinds: *building-up* or *constructive* processes, by means of which the raw materials of the food are converted into organic substances, which are incorporated into the fabric of the plant body or stored in the cells; *breaking-down* or *destructive* processes, during which substances of a more or less complex nature are decomposed into simpler bodies, energy being placed thereby at the disposal of the plant. In addition to the energy so obtained, green plants have at their disposal the vast stores of energy represented by sunlight.*

Constructive changes have been included under *assimilation* in its widest sense: assimilation of mineral salts and of nitrogen, and assimilation of carbon, or photosynthesis. *Destructive* changes have been considered under *respiration*.

In order to study these different aspects of nutrition, it has been necessary to treat them as if they were distinct and separate processes taking place independently in the plant. So far from this being the case, we know that constructive and destructive chemical changes are going on simultaneously in the living cells of the plant. Some of the reactions so involved do not occur in the laboratory, or can be effected by the chemist with difficulty. Examples of such are the syntheses of carbon dioxide and water to form sugar, the conversion of sugar to starch, and the alteration of starch to oil,—all of which reactions take place in plant cells at the ordinary temperature.

In the laboratory, chemical changes may take place spontaneously if certain substances are brought into contact, or they may be initiated or accelerated by heat. Sometimes they are brought about by the use of bodies known as *catalysts*, which accelerate, or appear to initiate, changes in substances with which they are in contact, although not themselves affected. In the living cell, chemical changes are usually effected by the action of *enzymes*, which may be considered as *organic catalysts*. So important is the action of enzymes that an understanding of the details of plant nutri-

---

* The changes taking place during nutrition of plants and animals are included under the general term *metabolism;* those of a constructive nature are described as *anabolic* changes, those of a destructive kind as *katabolic* changes.

tion can almost be identified with a knowledge of enzymes and their behaviour. It is one of the most wonderful facts in the physiology of plants that complicated chemical reactions of very different or even opposite character, brought about by the agency of different enzymes, can take place simultaneously within the limited compass of a minute vegetable cell.

Practical Work

1. **Diastase**—Diastase is present in large quantities in germinating seeds. An active extract can be prepared as follows. Germinate a quantity of barley in a germinator or on blotting-paper between two dishes. When well germinated, pound the seeds and seedlings in a mortar with a little sharp sand and shake up the pounded mass in distilled water. Filter. To the resulting clear filtrate add alcohol until a white precipitate is formed. Collect the precipitate on a filter paper, wash with alcohol, and redissolve in water. (The precipitate is not a pure substance chemically, but it contains the enzyme and is free from sugar.)

2. **Properties of Starch**—Take a *little* starch in a test tube with water. Boil. Note the behaviour of the starch. Cool: add a few drops of iodine solution. Note the characteristic colour reaction. Note also that the blue colour disappears on heating, since the compound of starch and iodine is decomposed at the higher temperature.

3. **Conversion of Starch**—Take a little starch boiled with water as described in the preceding paragraph. (The "starch solution" should be extremely weak.) Add to it, in a test tube, a little of the solution of diastase (experiment 1) and stand for twenty minutes or half an hour in a warm place (a warm room or in a water bath at a temperature not over 50° C.).

Test for sugar (glucose) as follows. Add a little *Fehling's solution* to the solution. Heat. If a "reducing" sugar* is present, the blue colour of the solution disappears and a yellow-red precipitate is formed as the solution approaches boiling point. N.B. The appearance of the precipitate *after some minutes' boiling* is not a satisfactory proof of the presence of a reducing sugar.

In order to convert the whole of the starch present, it is necessary to arrange the experiment so that the sugar will diffuse away as it is formed (p. 121). When this is done, by placing the starch and diastase solution in a "dialyser" or a bag of parchment membrane

---

* Many different sugars are known to the chemist. Certain of them are known as "reducing sugars" because they have the property of "reducing" an alkaline solution of a copper salt and certain other solutions when warmed with them. Among such sugars are malt sugar (maltose), grape sugar or glucose (dextrose) and fructose (levulose) all of which are formed by the action of diastase on starch in plants.

When copper sulphate in solution undergoes "reduction" the solution becomes colourless and a reddish precipitate of cuprous oxide appears.

the failure of the iodine reaction should be noted as the starch disappears owing to its conversion to sugar. The colour is at first blue, then purple, changing to reddish-brown and finally disappears. These colour changes mark the formation of intermediate substances, as the hydrolysis of starch proceeds.

4. Repeat experiment 3, using a few drops of concentrated sulphuric or hydrochloric acid instead of the solution of diastase. Boil; test for the presence of a reducing sugar as before.

5. Repeat experiment 3, using a little *saliva* instead of the solution of diastase. From the result, infer the presence of diastase in the saliva of human beings.

6. Repeat experiments 3 or 5, but boil the solution of diastase or saliva before adding to the starch solution. Infer the effect of heat on enzymes.

7. The effect of diastase on starch can also be studied by cutting thin sections through the storage regions of various seeds, e.g. Pea, Bean, Wheat. Examine microscopically and identify the starch by means of iodine. Treat the sections with a drop of the diastase solution, lay in a warm place and note the behaviour of the starch grains. Examine, also, a little of the reserve material from a Wheat seed some time after germination and note the condition of the starch grains.

8. **Proteases**—The action of *protein-splitting* enzymes can be studied by using a commercial preparation of *pepsin* (prepared from animal tissues) and observing its action on insoluble proteins, e.g. small fragments of boiled white of egg or the aleurone grains of seeds.

9. **Zymase**—For the action of zymase in Yeast, compare experiment 1, Chapter II.

PART II—REPRODUCTION

CHAPTER IX

REPRODUCTION

REPRODUCTION: GENERAL ASPECTS OF; VEGETATIVE AND SEXUAL REPRODUCTION. REPRODUCTION OF CELLS: NUCLEAR DIVISION AND CELL DIVISION. FORMATION AND DIFFERENTIATION OF TISSUES. VEGETATIVE REPRODUCTION: ZOOSPORES; SPORES AND SPORE FORMATION IN VARIOUS GROUPS OF PLANTS; GEMMÆ. POLLEN GRAINS OF SEED PLANTS COMPARED WITH SPORES. SEXUAL REPRODUCTION: FORMATION OF GAMETES; FERTILISATION; THE ZYGOTE. ORIGIN OF SEXUAL REPRODUCTION AND ITS RELATION TO VEGETATIVE OR ASEXUAL REPRODUCTION. SEXUAL REPRODUCTION IN BROWN SEAWEEDS. EVOLUTION OF SEXUALITY. MEIOSIS. ALTERNATION OF GENERATIONS

LIVING organisms, having reached maturity, tend to reproduce, i.e. to give rise to new individuals. The ways in which reproduction takes place are manifold, but they can all be referred to two types, vegetative and sexual, both of which are represented among plants.

Vegetative Reproduction—New individuals may be produced by division, equal or unequal, of the body of the parent: if the division is equal, the body of the parent divides into two parts approximately or exactly alike; if it is unequal, a fragment of the body, large or small as the case may be, is detached: in either case a new individual results. The formation of new individuals in such a manner is called *vegetative reproduction*, because the growing or vegetative body of the parent plant or animal is responsible for it.

Sexual Reproduction—In this type of reproduction, the formation of a new individual needs the co-operation of two parent cells called the *sex-cells* or *gametes*. In the case of unicellular plants, the ordinary vegetative cells behave as gametes, fusing in pairs to give rise to new individuals, which may persist as such or divide later to form several plants.

In multicellular plants, special cells are produced which function as sex-cells or gametes. The fusion of gametes which precedes sexual reproduction, whether in plants or in animals, is called *fertilisation*, and after fusion such cells are said to be *fertilised*. In all cases, the product resulting from fusion of two gametes is called a *zygote* (Greek, ξυγὸς—a yoke). This method of forming a new individual is called *sexual reproduction*, because it involves the formation by the parent plant or animal of special cells, the sex-cells or gametes, to which is entrusted the work of reproduction.

Sexual reproduction is found in the vast majority of plants and animals. In some of the simpler green plants, e.g. the Seaweeds, there is evidence that it has been derived from vegetative reproduction during the course of evolution by the specialisation of cells, detached by the parent plant for the purpose of reproduction. In the higher plants both vegetative and sexual reproduction are often carried out by the same individual, but in the higher animals the power of reproducing the whole animal by vegetative means is absent. Each method of reproduction has advantages and disadvantages which will be discussed when we have learned the nature of the processes involved.

Since reproduction of any kind, whether by vegetative or sexual means, involves multiplication of cells, we must first consider how cells reproduce or multiply, remembering that in one-celled plants they separate and become new individuals, whereas, in the higher plants, they remain attached and so build up the many-celled body of the individual. There is reason to believe that many-celled plants have been derived from one-celled plants as the outcome of a tendency for new cells to remain attached instead of separating. In the early stages of evolution, such groups of cells were loosely connected and of the nature of colonies, but, owing to *physiological division of labour* among the cells of the colony, and the *differentiation of structure* incidental to this, the different kinds of cells in a colony became indispensable to one another and to the colony of which they formed part. This tendency led in time to the evolution of multi-

cellular plants with their highly complex bodies composed of
different kinds of cells and tissues, each carrying on a different
part of the life-work of the plant and each possessing a
structure specially suited to that purpose.

Cell-division—A typical plant cell, as we have already
learnt, consists of a small fragment of protoplasm containing

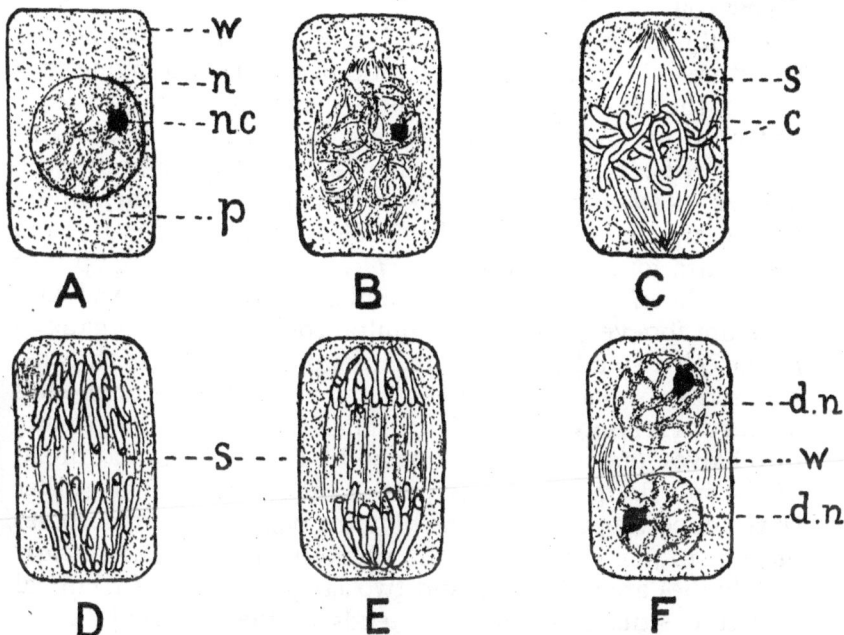

FIG. 20—KARYOKINESIS AND CELL-DIVISION IN A VEGETATIVE
PLANT CELL

A, young cell with resting nucleus.   B, nucleus preparing for division;
formation of spireme and disappearance of nuclear membrane.
C, chromosomes on equatorial plate of spindle.   D, chromosomes
(after splitting) passing to poles of spindle.   E, chromosomes
at poles.   F, division of nucleus complete; formation of new cell-
wall

w, cell-wall; *n*, nuclear membrane; *nc*, nucleolus; *p*, cytoplasm;
c, chromosomes; *s*, spindle fibres; *d.n*, daughter nucleus

a nucleus, plastids and other specialised structures, surrounded
by a cell-wall.   Increase in number is the result of division
of individual cells and does not take place directly but is
preceded by division of the nucleus, so that the whole process
can be studied in two stages: (1) division of the nucleus,
(2) division of the cell.

9

The nucleus may divide by simple constriction, thus separating into two or more pieces. This direct division or *fragmentation* of the nucleus does not occur in young, healthy cells of the growing regions, where it is replaced by a more elaborate process which results in an exact halving of the nuclear material, each half forming the nucleus of a new cell.

A cell-nucleus in the " resting " condition has a complicated structure. It contains a network of specialised living substance, the *chromatin ;* one or more dense bodies, the *nucleoli;* vacuoles containing a clear liquid, the *nuclear vacuoles* with *nuclear sap ;* and is surrounded by a delicate membrane, the *nuclear membrane.* When about to divide, the membrane disappears, the nucleoli and vacuoles lose their identity, the chromatin forms a thread, the *spireme,* which breaks up into pieces, the *chromosomes,* the number of which is fixed and constant for every species. Simultaneously, there is organised in the cell a spindle-shaped structure of protoplasmic threads, the *nuclear spindle,* at the centre of which the chromosomes arrange themselves. Each chromosome splits lengthwise· into halves, one of which is carried by contraction of a thread of the spindle to one end or pole of the spindle and the other to the opposite pole. On reaching the poles, the chromosomes once more join end to end, the vacuoles, nucleoli and nuclear membranes are re-formed, and two *daughter nuclei* of identical structure, situated at opposite ends of the cell, replace the original nucleus. This elaborate process, resulting in an exact division of the material of the nucleus, and in its distribution to two daughter nuclei, is called *indirect division, mitosis,* or *karyokinesis.* Nuclear division may be followed at once by *cell-division :* the cytoplasm forms a new cell-wall midway between the two nuclei, in the position occupied by the middle part of the spindle during nuclear division, and reproduction of the cell is complete. In unicellular plants the two new cells separate and form two new individuals; in multicellular plants they remain attached and help to build up a *tissue.* In certain special cases division of the nucleus is not followed by cell-division, or the latter is deferred until a number of nuclear divisions have taken place.

Plastids and other specialised portions of the protoplasm also undergo a process of division into two parts usually by constriction, such divisions taking place independently of cell-division.

Reproduction of Tissues—After the earliest stages of development, cell-division in multicellular plants is restricted to certain regions, thus composing a special kind of tissue called *meristem*. Growth in size of an organ may be regarded as taking place in two stages, (1) increase in the *number* of cells composing it, (2) increase in *size* of the additional cells. In Flowering Plants, growth in length of the stems and roots is due to the activity of such a tissue at or near the tips, the *apical meristem*, the cells of which are continually dividing, thus contributing to the formation of new tissues, a process which continues so long as the stem or root is capable of growth in length. Elongation of leaves takes place at first in the same way, but growth is of *limited* and not of *indefinite* duration as in the stem or root.

The organs of a flower are formed in a similar way, and, like leaves on a stem, are usually produced in a regular order, so that the youngest is nearest the tip of the stem from which they arise. Growth in thickness of stems and roots which continue to increase in girth, i.e. those of Dicotyledons and Conifers, is due to the activity of a layer of meristem situated between the wood and bast, the *wood cambium*, the cells of which divide continually during the growing season, thus adding to the bulk of wood and bast tissues available for the work of conduction. The *cork* which replaces the epidermis of young, unthickened stems is due to the activity of another layer of meristem, the *cork cambium*, the cells of which divide parallel to the surface, thus forming layers of cork which protect the stem from injury, or from excessive loss of water. In the mature tissues, groups or layers of meristem cells are also responsible for the formation of the *lenticels* or breathing pores which penetrate the bark of trees and shrubs, for the development of lateral roots, for the production of roots and buds on cut surfaces as in " cuttings," etc., and for the healing of wounds.

Such a dividing tissue, when composed of juvenile cells, is called a *primary meristem*. Examples of such are the apical meristems of the roots and shoots of Flowering Plants. In certain cases, cells which have become, in a sense, mature and lost their power to divide, regain their youthful character in this respect, and thus form a *secondary meristem;* of this nature are the cork cambium and parts of the wood cambium of roots and shoots.

A typical meristem, such as that found at the root tip of a Flowering Plant, is composed of young, unaltered cells of about equal diameter in all directions, fitting closely together. The cavity of each cell is filled with finely granular cytoplasm, in which is embedded a large nucleus. Bounding the cytoplasm externally and separating cell from cell is a delicate wall of cellulose.

Differentiation of Tissues—All the tissues of a Flowering Plant, and indeed of all plants, originate as cells of this kind; the younger cells of a meristem continue to form new cells by division, the older cells *differentiate* in various ways to form the tissues of the plant. Owing to their mode of origin, the cells of a tissue are older as they are more remote from the meristem which produced them. In a vertical section through, for example, the tip of a stem or root, the development of all the cells which compose the mature tissues can be traced from the tip backwards. At the tip they have the structure described above, a short distance back they lose the power to divide and begin to differentiate; some increase in size equally in all directions; vacuoles appear and may merge to form a single large vacuole, but the cells retain their nuclei and continue to be capable of growth and other activities.

Much of the soft, succulent part of a Flowering Plant is composed of such cells, which form a widely distributed tissue known as *parenchyma*. These cells separate from one another at the corners to form small triangular spaces which communicate with others of like origin to constitute a continuous system of *intercellular spaces*. On the outside of the young root, parenchyma cells grow out on one side to form root-hairs, thus constituting the piliferous layer of the root. The cells of the outer layer of the young stem and leaves complete their differentiation by undergoing a modification of the outer cell-wall; the thin cellulose wall of a parenchyma cell is extremely permeable and offers no protection from loss of water by evaporation from the cell vacuole. In the cells of the epidermis a substance named *cutin* is formed and excreted into the outer walls of the cells. When the cells of the epidermis are mature, cutin has accumulated in and upon the outer part of the wall, so forming a layer impervious to water called the *cuticle*.

In the more central parts of the roots, stems and leaves strands of cells differentiate to form the *conducting strands* or *vascular bundles*.

The cells of these strands are recognisable a short distance from the tips by reason of the proportionately greater increase in length of many of them. These elongated cells undergo further differentiation to form the *tracheids* of the wood and the *sieve tubes* of the bast: in Dicotyledons, rows of such elongated cells fuse end to end to form the *vessels* of the wood. Most of the tissues found in the vegetative parts of a Flowering Plant when mature can be referred to one of the three following groups.

1. *The Tegumentary Tissues*—Those found covering the surface of organs. In the young root this is an absorbing layer, the *piliferous layer* with *root-hairs;* in the young stem and leaf it forms a protective layer, the *epidermis*, replaced in older stems and roots by *cork* or *bark*.

2. *Conducting or Vascular Tissues*—Continuous strands or tracks traversing the roots, stems and leaves, serving as a main highway for the passage of water and food materials to and from all parts of the plant.

3. *Ground Tissue*—Composed largely of *parenchyma*, the cells of which can, not infrequently, regain their power to divide and so give rise to a local *meristem*.

In many parts of the plant, parenchyma cells of the ground-tissue become extremely thick-walled and give rise to rigid strands or layers,—the *skeletal* or *mechanical* tissues. In the leaf and the green parts of the stem, the cells of the ground-tissue contain chloroplasts and form the assimilating tissues of the plant (p. 85).

No attempt will be made to study in greater detail the complex anatomy of a Flowering Plant. The purpose of this brief summary will have been served if it has made clear to the student how reproduction of a plant cell takes place and how tissues originate.

Vegetative Reproduction—In unicellular plants, cell-division and vegetative reproduction mean the same thing, i.c. new individuals arise by division of the ordinary vegetative cells.

In multicellular plants, vegetative reproduction often takes place by the separation of special cells, or small groups of cells, called *spores*, which develop into a new individual. Spore-production is a characteristic feature of the plants of many groups of the Vegetable Kingdom. Among the lowest green plants, or Algæ, which live in fresh or salt water or in very damp places, the spores are set free in the water and have a structure suited to their environment. Each spore is a cell, more rarely a group of cells, surrounded by a thin skin or pellicle formed by the protoplasm, but not by a cellulose wall. Embedded in the protoplasm of each cell is a nucleus, often one or more plastids, in addition to which other specialised structures may be present. These spores swim freely in the water by means of delicate threads of protoplasm, the *cilia* or *flagella*, which project from the surface of the cell and execute characteristic movements in the water. A similar power of moving through water is shown by many minute animal forms of life, for which reason, when first observed, these motile spores were named *zoospores* (Greek: ζῷον=an animal).

Zoospores are also formed by a few species of fungi which grow in water or damp places, but the majority of these plants live on dry land and produce spores adapted for dispersal in the air. Such spores are small and light and are protected from desiccation by a cell-wall. The degree of vitality they possess is very variable; the wall may be very thin and such spores quickly perish, unless they find a place sufficiently moist to allow them to germinate at once; other spores have thick walls which protect them from injury and allow them to remain in an inactive or resting condition for a considerable time, until conditions suitable for germination present themselves. Minute spores are constantly liberated into the air by common fungi, such as the "moulds" which grow on bread, cheese, or jam. If such materials are left about uncovered, they serve as a kind of "seed-bed" on which spores, floating in the air, settle down, germinate, and give rise to another colony of the fungus or "mould" to which they belong. The spores of fungi or other lowly organisms which live as parasites may also be present in the air and may enter the body of a suitable animal or plant, to germinate there and eventually set up the symptoms of disease.

There is wonderful diversity in the structure of spores and the manner in which they are formed: sometimes they are cells, detached singly or in groups from some part of the plant; sometimes they are produced by cell-division in the contents of a special organ or *sporangium*, the wall of which splits to liberate them when ripe (fig. 21).

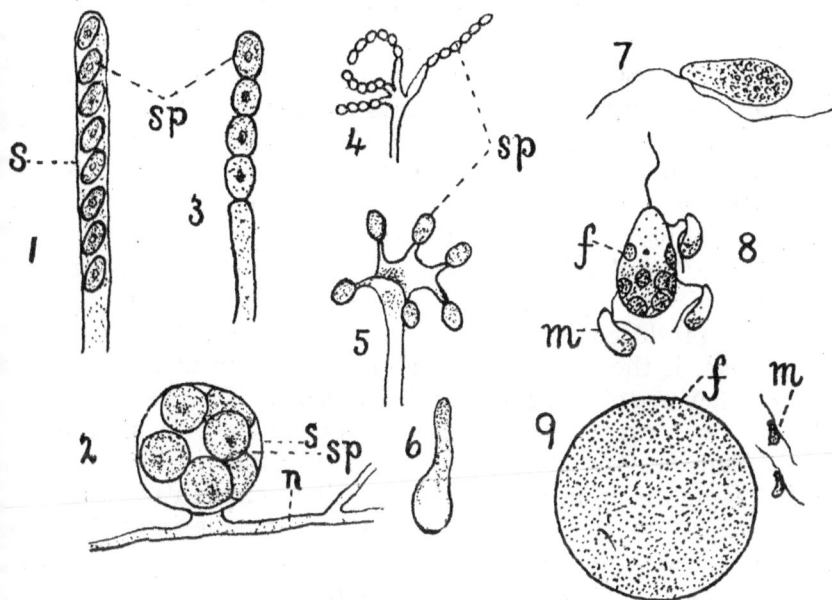

FIG. 21

1 and 2, Endospores of Fungi; 3, 4 and 5, conidia of different species of fungi borne on variously shaped aerial branches; 6, germinating fungus spore; 7, 8 and 9, sex-cells or gametes of Brown Seaweeds; 7, male gamete of *Cutleria multifida* ; 8, male and female gametes of *Ectocarpus secunda;* 9, male and female gametes of Fucus sp.

s, sporangium; *sp*, spore; *n*, vegetative hypha; *m*, male gamete; *f*, female gamete

The Mosses and Liverworts produce their spores in large sporangia or *sporogonia* which are often a conspicuous feature of the plant.

Some such plants, in addition to or in place of spores, produce larger structures called *gemmæ* each of which when detached grows into a new plant. Of this nature are the small green bodies produced by some common Liverworts and Mosses.

In Ferns and their allies, spores are also produced in sporangia, grouped together on some part of the leaves or fronds.

In Seed Plants generally,—the Yew, and its allies the cone-bearing trees (Gymnosperms), and the Flowering Plants (Angiosperms),—the pollen grains and certain structures within the ovules are, *strictly speaking*, of the nature of spores, which behave like those of a fern and give rise to male and female gametes (p. 139).

Sexual Reproduction—The essential feature of sexual reproduction is the formation of *sex cells* or *gametes*, which fuse in pairs to form a cell of dual origin, the *zygote*, which develops into a new plant. When unicellular plants reproduce in this way, the whole plant behaves as a gamete, and fuses with another individual of the same species to produce a zygote which may develop directly into a new plant or may give rise to several new individuals by subsequent division. The gametes produced by multicellular plants may be alike or dissimilar; if the former, there is no distinction of *sex*, if the latter, they are of two sexes, *male* and *female*. If the gametes are of two sexes, they may be produced by two separate individuals or in different parts of the same individual.

In plants, as in animals, it is difficult to explain the *origin of sex*—i.e. how the differences arose among the individuals of one species which we call a difference of sex. There are, however, many facts to be gleaned from a study of sexual reproduction in the simpler plants which help us to understand how it may have come about, and the manner in which it is related to vegetative reproduction.

In certain of those common plants of the sea shore the Brown Seaweeds, vegetative reproduction takes place by the formation of a number of separate cells or zoospores, which escape into the water and swim about as independent organisms. After a time, they settle down on a rock or other solid body, and give rise by cell-division to new plants.

Among the Brown Seaweeds, it is not uncommon for a single species to reproduce asexually by means of zoospores and sexually by means of gametes. When the latter occurs, gametes are set free in the water and are often identical in appearance with zoospores. They behave, however, as gametes—i.e. they cannot give rise to new individuals without fusion,

The gametes produced by some Brown Seaweeds are all alike; in other species, they are of two sizes, large and small, and a zygote is formed only when a large female cell fuses with a small male cell. Another Brown Seaweed, the common Bladder Wrack (*Fucus vesiculosus*), possesses two kinds of gametes, markedly different both in structure and in behaviour, produced in different organs of the same parent plant. In one kind of organ (*antheridium*) there are produced small active male cells (*antherozoids*), which swim about freely in the water; in the other kind of organ (*oögonium*) are formed a smaller number of large female cells (*ova, egg-cells*), containing a quantity of reserve food materials and incapable of movement.* Both male and female gametes are set free in the sea-water in which the plants are growing, and a zygote is formed when a small active male gamete fuses with a large passive female gamete or egg-cell. In other plants of the same group, differentiation of sex has advanced a step further; certain individuals of a species produce male gametes only,—they are of the *male sex ;* other individuals produce female gametes or egg-cells only,—they are of the *female sex* (fig. 21).

The interpretation of these facts seems to be that sexual reproduction in the Brown Seaweeds originated as a special modification of vegetative or asexual reproduction, in consequence of the inability of certain asexual reproductive cells to grow into new plants unless they had previously fused in pairs. The principle of " division of labour," always at work in living things, whether among the individuals of a colony or society, or among the cells of an individual, led subsequently to the production of two kinds of reproductive cells, one kind small and active, the other kind larger and less active owing to the storage of food reserves—in other words, to the differentiation of male and female gametes. The same principle, operating among the individuals of a species, led to the evolution of plants which produced only male or female gametes respectively, and thus, to *differentiation of sex* in individuals. It is not improbable that a similar interpretation can be

* The inclusion of large stores of reserve material in the female gamete or egg-cell, for the benefit of the young plant or animal which will arise from it, leads frequently to the formation of egg-cells of large size. This is notably the case in the eggs of birds, which may be of great size owing to the reserves stored in the yolk and white.

applied to the existence of two different sexes in other groups of plants and in animals.

Meiosis (μείωσις, reduction)—The number of chromosomes appearing at nuclear division in the cells of plants and animals can be counted, and is always the same for any one species. Thus, the vegetative cells of a Lily have 12 chromosomes, those of Wheat have 16 chromosomes; there are as many as 168 chromosomes in the cells of some animals, while there are only 2 or 4 chromosomes in the thread-worm *Ascaris*.

If all nuclear divisions were of the type already described, fusion of gametes at sexual reproduction would involve doubling of the number of chromosomes characteristic of the species, and the zygote would always possess twice as many chromosomes as did the cells of the parents. Moreover, this increase in number would be repeated in each generation, so that the number of chromosomes in the cells of successive generations of an animal or plant reproducing by sexual methods would show a steady increase in geometrical progression. This increase in number does not take place and is avoided in the following remarkable way.

The gametes or sex-cells of an animal or plant always contain *half* the number of chromosomes characteristic of the species, the reduction in number taking place during a special cell-division which precedes the formation of sex-cells. Thus, while the vegetative cell of a Lily contains 12 chromosomes, each sex-cell contains only 6 chromosomes. This special division of the nucleus is called the *reduction division* and the stage at which it occurs is called *meiosis*.

In animals, a reduction division immediately precedes the formation of gametes; in the Vegetable Kingdom this is rarely the case and it occurs at a different stage in the life history in different groups of plants. Thus, in the Bladder-wrack (*Fucus vesiculosus*) it immediately precedes the formation of the sex-cells, as in animals. In other species of Algæ this may or may not be the case. In Mosses and Liverworts, reduction takes place at the formation of the spores, and each spore possesses only half the number of chromosomes present in the cells of the sporogonium. The same is true of Ferns and their allies, and, since in this group of plants, the spore on germination does not give rise directly to a new fern plant but to a small independent green structure called the *prothallus*, certain cells of which produce the gametes, we have a striking manifestation of the remarkable phenomenon known as *alternation of generations.*

In the Higher Plants a reduction division takes place both in the stamens and in the ovules. In the former, it immediately precedes the formation of pollen grains (*microspores*); in the latter, it takes place in a cell of the *nucellus* of the ovule, and results in the formation of a special cell called the *embryo-sac* (*megaspore*). The male and female gametes, produced respectively by the pollen grain and the embryo-sac, each carry the reduced number of chromosomes and, by their fusion, the full number is restored at fertilisation.

The course of events during a reduction division may be briefly described as follows.

The preliminary changes which take place in the nucleus prior to the appearance of chromosomes differ in several particulars from those of ordinary cell-division, and result in the appearance of *half* the typical number of chromosomes. Each of these chromosomes, although apparently single, consists, in all probability, of two ordinary chromosomes joined side by side or end to end. At the splitting of chromosomes which precedes their distribution to the two ends or poles of the spindle (p. 130), the two constituents of each "double" chromosome separate. A reduction division results, therefore, in an equal distribution of the original chromosomes to the two daughter nuclei. At the next division, each of the whole chromosomes splits in the usual way; the products of division, therefore, are cells with chromosomes of the ordinary kind, but of which only half the number are present.

Alternation of Generations—The life history of a Fern consists of two successive stages or phases, one of which, the *fern plant*, possesses cells with the full number of chromosomes characteristic of the species. Preceding spore-formation there is a reduction division, as a result of which the nucleus of each spore contains only half the number of chromosomes, and this reduced number is found in each cell of the prothallus which results from germination of a spore. The gametes, which originate from special cells of the prothallus, thus carry the reduced number of chromosomes and, by their fusion at fertilisation, form a zygote, and subsequently a new fern plant with the full number of chromosomes in each cell.

This alternation of a *spore-bearing phase* (*sporophyte*) with a *gamete-bearing phase* (*gametophyte*), the latter consisting of cells with the reduced number of chromosomes, is a widespread phenomenon, but it is only rarely that, as in ferns, the two stages are represented by distinct and independent plants.

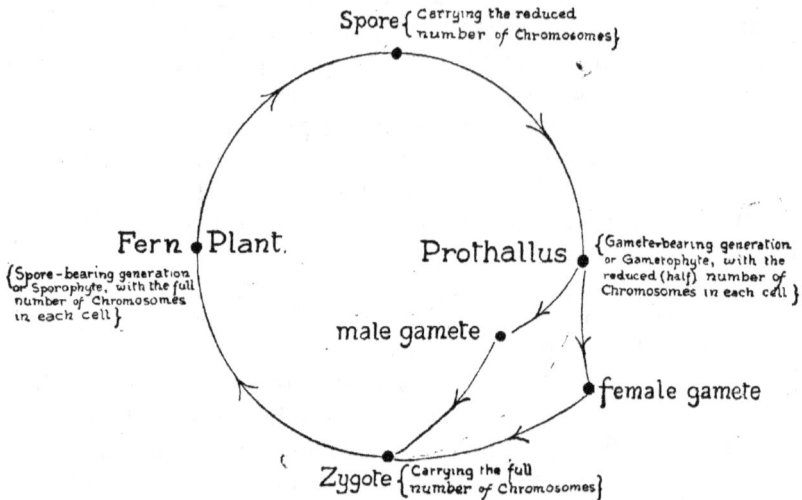

Spore { Carrying the reduced number of Chromosomes }

Fern ● Plant. {Spore-bearing generation or Sporophyte, with the full number of Chromosomes in each cell }

Prothallus ● {Gamete-bearing generation or Gametophyte, with the reduced (half) number of Chromosomes in each cell }

male gamete ●

● female gamete

Zygote { Carrying the full number of Chromosomes }

FIG. 22—DIAGRAM ILLUSTRATING THE LIFE CYCLE OF A FERN

Much discussion has centred around the nature of " alternation of generations " in plants, and there is still some disagreement among botanists as to the significance of the two phases. With this aspect we are not concerned; the interest of the subject for us lies in the methods by which reduction in number of the chromosomes is effected, and its possible bearing upon problems of inheritance (p. 180).*

* It should be noted that the presence of a different number of chromosomes in the nuclei of the sporophyte and gametophyte is not necessarily correlated with a difference of form in the two phases. In the Brown Seaweed, *Dictyota*, for example, the two phases are identical in appearance, although the cells of one contain twice as many chromosomes as the cells of the other. Apart from the number of chromosomes, the plants can only be distinguished by the fact that spores are borne by the one and gametes by the other. Presumably the characteristic forms of sporophyte and gametophyte in plants do not depend upon the number of chromosomes contained in their nuclei, although the half number is usually associated with the gametophyte and *vice versa*. The essential point is that " reduction " takes place at some stage in the life history of all plants which reproduce by sexual means

Practical Work

1. If suitable microscopic preparations and a microscope with high-power lenses are available, slides can be examined showing nuclei at various stages of division; also longitudinal sections through the apices of stems and roots to show the apical meristems and region of differentiation. Sections of stems and roots can also be examined in order to illustrate the different tissues mentioned in the preceding chapter. It does not come within the scope of this book to describe this work in detail and it is assumed that such laboratory exercises have been, or will be, carried out by the student in the course of his botanical studies.

2. Collect the fine dust-like spores produced by a mould by suspending a piece of mouldy bread or other material in still air under a bell-glass or beaker and collecting the spores upon a piece of dark paper or card. In a similar way good *spore-prints* can be got by cutting a fully developed mushroom at the top of the stalk and placing it, gills downward, on a sheet of white paper. Spore-prints of other " toadstools " belonging to the same group as the mushroom, but with white or pale-coloured gills, can be made by placing the spore-bearing parts on *black* paper or card.

The spore-cases of mosses and ferns should also be examined and the presence of spores determined by means of a hand lens or by laying the spore-bearing parts on white paper. If a microscope is available spores of fungi, etc. can be examined dry or mounted in water. For germination of spores see experiment 5, Chapter X.

CHAPTER X

REPRODUCTION (*continued*)

REPRODUCTION IN SEED PLANTS: VEGETATIVE REPRODUCTION BY
RHIZOMES, TUBERS, ETC.; ARTIFICIAL METHODS OF VEGETATIVE REPRO-
DUCTION. SEXUAL REPRODUCTION IN FLOWERING PLANTS: COMPARATIVE
ASPECT. SIGNIFICANCE OF THE FLOWER. SEED FORMATION IN RELATION
TO TERRESTRIAL HABIT. THE STAMENS; POLLEN. THE CARPELS;
STRUCTURE OF OVULES. POLLINATION. FERTILISATION. DEVELOP-
MENT OF SEED AND FRUIT. PARTHENOGENESIS. "RIPENING" OF
SEEDS. VITALITY OF SEEDS. "GERMINATION CAPACITY." ADVAN-
TAGES OF THE "SEED HABIT." VEGETATIVE AND SEXUAL REPRODUC-
TION: GENERAL SUMMARY

Vegetative Reproduction in Flowering Plants—Among
Flowering Plants, vegetative reproduction takes place by the
separation of larger and more highly differentiated structures
than the spores and gemmæ produced by the lower plants.
Vegetative reproduction is common among herbaceous
perennials, in which the purpose is often served by specially
modified portions of the shoot system which function both as
storage organs for food material and as a means of reproduction. Of this nature are the rhizomes of *Iris*, Solomon's
Seal and many Grasses, the tubers of Potato and Jerusalem
Artichoke, the corms of *Gladiolus* and *Crocus*, the bulbs of
Narcissus, Tulip, Snowdrop and many other plants. All of
these are underground shoots, the tissues of the stems or
leaves of which contain an abundant store of starch or other
reserve foods. Some plants form buds upon the roots, either
spontaneously or if injured: from such buds originate the
suckers which spring from the roots of Elm, Poplar, and Rose;
the swollen tap roots of the Dock and Dandelion can behave
in the same way, a fact which must be taken into account
when attempting to rid the garden of these plants; gardeners
turn the same property to advantage when they propagate
Seakale and other plants by means of root-cuttings.

The *runners* of the Strawberry and the *offsets* of the House-leek serve a like purpose, and are largely made use of in the cultivation of these plants. In yet other plants, special buds or *bulbils*, often swollen by the storage of reserves, are detached by the plant and function as a means of vegetative reproduction. Of this nature are the bulbils on the stems of certain Lilies and on the margins of the leaves of *Bryophyllum*, while the gardener takes advantage of the formation of buds upon the pegged down leaves of *Gloxinia*, *Begonia* and many ferns in order to propagate them. Various methods of propagating plants artificially have been devised by gardeners and cultivators. Many plants will grow from *cuttings*, portions of the shoot severed close to a node, which form roots upon the cut surface if placed under suitably moist, warm conditions. Other plants can form roots upon the injured surfaces if portions of the shoot are cut partly through and pegged-down below the surface of the soil, or otherwise kept moist; Carnations and certain trees and shrubs are often *layered* in this way.

The *budding* or *grafting* of roses or fruit-trees depends upon similar powers of regeneration in cut surfaces, and the significance of the layer of meristem called the wood cambium should be noted in this connection (p. 131).

Such are a few examples selected from the almost innumerable cases of vegetative reproduction, natural or artificial, found among Flowering Plants.

Sexual Reproduction in Flowering Plants—Necessity for fusion is the characteristic feature of all gametes. In the lower green plants or Algæ these cells are set free from the parents, and gametes of both kinds or of one sex only can move freely in the water in which such plants grow and so approach one another and fuse. This mode of reproduction evidently necessitates an aquatic or semi-aquatic mode of life. In contrast with the Algæ, the majority of Flowering Plants live on land, and produce gametes of two sexes neither of which have any power of free movement. Yet, for the production of a zygote,—the first stage in the development of a new plant,—a male gamete must fuse with a female gamete. The evolution of the means whereby this may be accomplished has led to many characteristic features in Flowering Plants; the manifold differences of shape, colour and structure in flowers, for example, are devices which assist indirectly towards its accomplishment.

Before describing the course of events in a Flowering Plant and comparing it with sexual reproduction of a simpler kind, certain facts must be noted.

In the strictly botanical sense, it is not correct to compare a Flowering Plant directly with a primitive green plant or Alga producing male and female gametes, nor is it strictly correct to speak of the organs of the flower or of the individual plant as being of the " *male sex* " or of the " *female sex* " according to the kind of gametes produced. To explain the reasons for this statement demands more space than we have at our disposal and needs a study of plants in general for which we cannot spare time.

This comparative aspect of plants, although of great interest to botanists, does not in any way affect the mode of behaviour of the gametes or the development of the zygote into a new plant and can therefore be temporarily ignored.

Regarded from the point of view of *function*, or the purpose which it serves in the plant, a flower is a *reproductive* as compared with a *vegetative* shoot. The gametes are produced by the stamens and carpels of the flower: the stamens produce a quantity of dustlike pollen, each grain of which contains two male cells or gametes; the carpels, singly or collectively, form the *pistil* of the flower, within a hollow portion of which, the *ovary*, are born small budlike structures, the *ovules*, each of which contains, when mature, a single *egg-cell* or female gamete. In order to produce a zygote, one of the male gametes must fuse with another minute speck of living material,—the egg-cell,—contained within an ovule, produced within the ovary of the same or of another flower.

The gamete of a Flowering Plant is a *cell*, i.e. it consists of a nucleus surrounded by cytoplasm, although the amount of cytoplasm in the male gamete may be extremely small. In neither male nor female gamete is the cell invested by a cellulose cell-wall. The result of fusion is a zygote which develops forthwith into a miniature plant or embryo within the ovule. This period of development is one of great activity in the pistil of the flower and puts a strain on the resources of the parent plant. Much food material is required in order that an embryo plant may be formed from each fertilised egg-cell or zygote, in addition to which a supply of reserve food material is accumulated in or around the embryo to serve later as a source of supply during germination and the early stages of growth.

During development of the embryo and storage of such reserves within the ovule, the latter undergoes great changes. It increases in size, much of the internal structure disappears and is replaced by the embryo plant and its associated food reserves. The outer covering of the ovule, comprising one or two coats or *integuments*, undergoes change, and the whole ovule, partly by alteration of portions, but in great part by the development of entirely new structures,—the embryo plant and its reserves,—becomes the organ which we call a *seed*.

The reserves may be stored in a special tissue around the embryo, the *endosperm*, in which case the seed is described as *endospermic*, or within the body of the embryo itself, when the seed is described as *non-endospermic*. In either case, the seed consists essentially of a new plant with a store of food-reserves, surrounded by a jacket,—thin, thick, leathery, or woody as the case may be,—the *testa* of the seed. Simultaneously with the development of the *ovules* into *seeds*, the ovary of the flower develops into the *fruit*, a change which not infrequently involves also persistence of other parts of the flower. The infinite variety of structure found in fruits and seeds and its relation to the multitudinous ways in which they are dispersed,—whether by wind, water, animals or explosive movements,—are matters of very great interest to the student of plants, and are dealt with at greater length in other textbooks.

Owing to the manner in which the gametes of Flowering Plants are produced, the processes which lead to their fusion . take place in *two* stages. The first stage, *pollination*, consists in transference of pollen from the stamen to a specialised part of the pistil, the *stigma*, of the same or another flower; the second stage, *fertilisation*, includes transference of a male gamete from a pollen grain to the neighbourhood of an egg-cell and its subsequent fusion therewith.

Pollination is effected by pollen reaching the stigma of the same flower or another flower. The marvellous diversity of structure in flowers; the colour and shape of the petals, production of nectar and scent, are closely related to the mode of pollination by wind, insects and other means, more especially to *cross-pollination*, or transference of pollen from one flower to another by insects.

Sexual reproduction in Angiosperms is thus a complicated affair, involving directly and indirectly a number of complex processes. To the student of evolution there is little doubt that the organ called in common language the *flower* has been evolved in relation to the nature of the gametes and their mode of fusion and that these, in turn, are directly related to a gradual and complete adaptation of plants to a terrestrial mode of life.* The Angiosperms represent the most advanced stage which has been reached by green plants along this line of evolution.

A similar statement to that just made concerning the flower may be made respecting the evolution of the *seed-habit*, and there is abundant evidence that the habit of forming seeds has been evolved side by side with the evolution of land plants from ancestors which were more or less aquatic in their mode of life, as are the simpler green plants,—the Algæ,—to-day.

Formation of seeds marks complete adaptation to life on dry land, and the seed is essentially a device by which the plant can safely tide over a critical stage in its life history. The embryo plant within a seed can remain alive through the most unfavourable conditions for growth, such as drought or cold, and is reawakened to activity at germination by just those conditions which are favourable to plant growth.

The arguments upon which these statements are based can only be understood by making a comparative study of plants in general, which enables us to trace the origin and development of sexual reproduction in seed plants and to link it with the corresponding process in other groups. It is only possible here to deal briefly with the mode of origin of the gametes and the manner in which they come together and fuse.

The Stamens—The young stamen grows out from the stem or axis of the flower in much the same way as does the leaf from

* The existence of Flowering Plants which are aquatic in habit does not affect the argument, since it can be readily demonstrated that such have become modified in relation to their special mode of life.

the stem of the leafy shoot. As it grows it usually becomes differentiated into a thicker terminal part, the *anther*, borne upon a more slender part, the stalk or *filament*. Hairs or appendages of various kinds may develop later upon both filament and anther. The pollen grains are formed from a special meristem,—the *archesporial tissue*,—lying under the outer layers of cells of the anther, and in many cases there are four strands or patches of such cells. Each cell of the archesporium has a large nucleus and dense protoplasm and is separated from its neighbours by a delicate wall of cellulose. As the flower approaches maturity these cells separate from one another and each divides twice in rapid succession to form four cells,—the young pollen grains. As a consequence, the region of the anther formerly occupied by archesporial tissue becomes filled with separate cells,—the pollen grains. Each cavity formed in the anther in this way is called a *pollen-sac* and, since in the majority of anthers there are four strands of archesporial tissue, so there are four pollen-sacs, from which the pollen grains escape subsequently by splitting of the anther wall.

There is much variety in size, shape, and colour of pollen grains. They may be spherical, flattened, oval, four-angled or polygonal; in a few plants they are barrel-shaped. In size, pollen grains also differ from one another greatly; those of a plant with very large grains may be 100 times as large as those of another plant with very small grains. In the former case there may be as few as 32 large grains in each pollen-sac of the anther; in the latter as many as 60,000 small grains.

In the Pine and many of its allies, each pollen grain has two little floats or air-sacs which serve to make it easily blown about by the wind.

In some plants the pollen grains remain attached together as formed in sets of four, e.g. in the Cranberry and the Bilberry (*Vaccinium sp.*); in others, the grains are surrounded by a sticky substance which pulls out into delicate threads when they separate, e.g. *Rhododendron*, or the Evening Primrose (*Œnothera biennis*).

The outer wall of a pollen grain is often thickened and may be spiny or sculptured in various ways; whatever its character in the ripe pollen grain, there are always present one or more thin places, the significance of which will appear later.

In the pollen grain of a Sedge there is only one thin place or pore in the wall; in that of the Evening Primrose there are three; in that of *Convolvulus* 15 to 18 pores. Sometimes the thin places have the form of little oval or circular windows; in a few plants they may have a more complicated

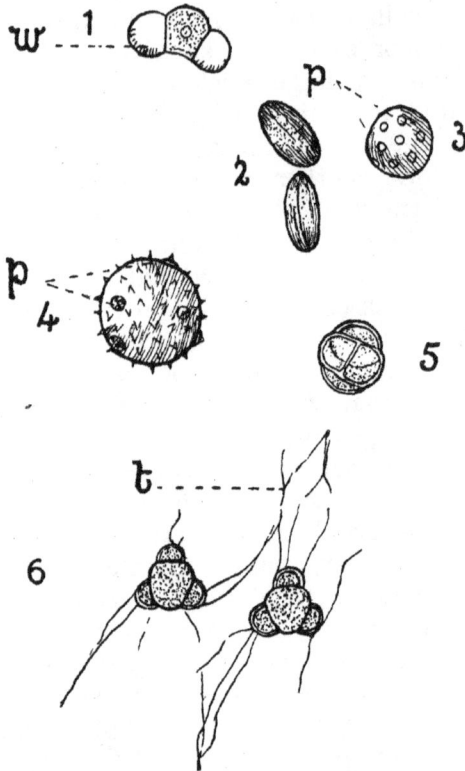

FIG. 23—FORMS OF POLLEN GRAINS

1, Pinus; 2, Gentian; 3, *Convolvulus;* 4, Cucumber; 5, Vaccinium (tetrad); 6, Evening Primrose, attached by threads of viscum

w, wing; *p*, pore; *t*, sticky threads

structure, as in the Vegetable Marrow, in which they are ring-shaped, and surround a small area of the thicker part of the wall, which thus forms a little lid (fig. 23).

When shed from the anther, each pollen grain contains three cells, i.e. three separate nuclei surrounded by protoplasm but not separated from one another by cell-walls.

The Carpels—The carpels may be produced in the same flower as the stamens, in which case they are the latest structures to be produced in the flower of which they form the central part. There may be one carpel only as in the **Pea** flower; more often there are several or numerous carpels, in which case each may form a distinct part of the pistil, or may fuse with the others to form a single structure. The essential features of the pistil of a flower are the formation of a closed structure, the *ovary*, within which the *ovules* arise, and the modification of the terminal part of the pistil to form a *stigma*, the surface of which is specially adapted for the retention of pollen grains which fall upon it or reach it in other ways. Each ovule is a small, stalked, bud-like structure originating from the inner wall of the ovary. The details of ovule development and arrangement of the ovules on the walls of the ovary will not be considered here. For our purposes it suffices to give a brief account of the structure of a typical ovule when mature.

A small oval body borne at the end of a short stalk, the *funicle*, the ovule projects into the cavity of the ovary or, more often, becomes curved during growth so that its tip or apex comes to be situated near the base of the stalk (see fig. 24).

The central portion, comprising the greater part of its bulk, is a mass of tissue called the *nucellus*, around which grow up one or two coverings or *integuments*, each composed of a number of layers of cells. The inner integument springs from the base of the nucellus and envelops it completely in a cup-like manner, except for a narrow channel at the apex where the edges of the cup come together. The second integument, when present, springs from below the first and completely envelops it except at the apex. The narrow passage between the two lips or edges of the integuments is called the *micropyle*, and it serves as the only means of communication between the cavity of the ovary and the nucellus of the ovule. Near the top of the nucellus, just at the base of the micropyle, a single cell increases considerably in size, so forming what is called the *embryo-sac* of the ovule.

The embryo-sac is at first a cell with a single nucleus, but it undergoes a series of divisions and in the mature ovule contains several cells,—eight in the majority of plants, a larger number in a few cases.   Two only of the structures within the embryo-sac are directly concerned with fertilisation and we will confine our attention to these.   The first and most important is the *egg-cell*, a large nucleus surrounded by cytoplasm situated at the top of the embryo-sac near the base of the micropyle; the other, a pair of cells lying near the centre of the sac, called the *polar nuclei*.

The protoplasm of the cells contained within it does not fill the cavity of the embryo-sac, the central part of which is a

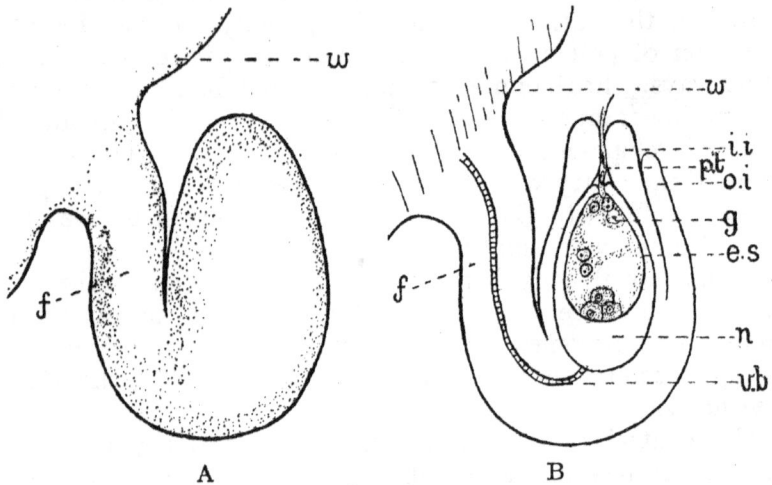

FIG. 24

A, ovule of Flowering Plant.   B, ovule in vertical section.   (Diagrammatic)

w, wall of ovary; *f*, stalk (funicle); *i.i*, inner integument; *o.i*, outer integument; *e.s*, embryo-sac; *n*, nucellus; *v.b*, vascular bundle; *g*, egg-cell (ovum); *p.t*, pollen tube

large vacuole filled with cell-sap.   This is the condition of the embryo-sac in the ovules of most plants at the time of pollination (see fig. 24).

Pollination—The pollen grains fall upon the stigma of the same flower,—*self-pollination*, or are carried to the stigma of another flower,—*cross-pollination*.   They adhere closely to the irregular surface of the stigma, which is often sticky owing to the exudation of a sugary liquid.

Fertilisation—Upon reaching the stigma the pollen grains behave in a very remarkable way.   They begin to germinate or grow by sending out delicate outgrowths or tubes, the *pollen-tubes*, into which the contents of the pollen grains pass. Each tube begins as a small bulge at one of the thin places or pits on the wall of the grain, and collectively they grow down through and between the cells of the stigma and through that part of the pistil called the style which connects the stigma with the ovary below.

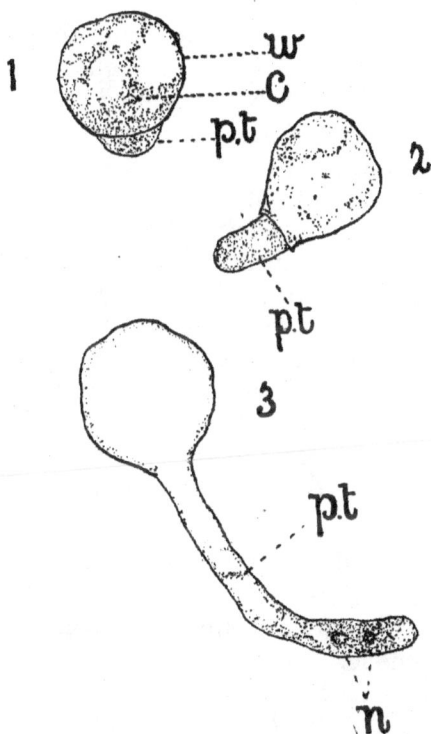

FIG. 25—STAGES IN THE GERMINATION OF POLLEN OF *Echeveria retusa* 1, after one hour; 2, after two hours; 3, after four hours; all in 15 per cent. cane sugar
w, wall of pollen grain; *c*, cytoplasm; *p.t*, pollen tube; *n*, nuclei

This growth of pollen-tubes through the tissues of the plant resembles that of a parasite which has invaded them, and the direction of growth is doubtless determined by the presence of sugar and other food materials within the cells of the stigma and style. The pollen-tubes reach the ovary chamber, grow around the walls, and each ultimately finds its way to the micropyle of an ovule, through which it grows.* The length of the pollen-tubes is determined by the distance of the stigma from the ovary of the flower, and the rate at which they grow varies from flower to flower. Many pollen-tubes may reach the ovary, but only one grows into the micropyle of each ovule.

During growth, the three cells which form the contents of the pollen grain pass into the pollen-tube and, at the moment when it grows into the embryo-sac of an ovule, they are found near the tip of the tube. Of these three cells, two are male cells or gametes, the other is probably related to the growth of the pollen-tube and plays no direct part in fertilisation. Within the micropyle, the delicate wall around the tip of the pollen-tube dissolves, and its contents pass into the embryo-sac. One of the male gametes fuses with the *egg-cell* and this fusion constitutes the act of fertilisation; the other male gamete has been observed in a number of cases to fuse with the *polar nuclei* (p. 150). Fertilisation results in the formation of a zygote which develops into a young plant; the second fusion of three nuclei results in the formation of a tissue called *endosperm*, the cells of which become filled with food materials to supply the growth of the developing embryo, and this endosperm tissue may or may not persist in the fully formed seed.

The Seed—The mode of formation of seed and fruit, subsequent to fertilisation, has already been briefly outlined (p. 145); a few words may be added dealing with some special features of seed development.

The later stages of development passed through by a seed whilst still attached to the parent plant involve certain remarkable changes in its constitution which are included under the term "ripening." Some of these changes concern the reserve materials, starch, oil, protein grains, etc., stored in the cells of the embryo or endosperm; the most profound and far-reaching are associated with the loss of a large proportion of the water present, with the consequent chemical and physical changes involved.

Whereas actively growing tissues, such as those of a de-

---

* In certain plants, the pollen tubes find their way to the embryo-sac by growing through the tissues of the nucellus at the end of the ovule remote from the micropyle.

veloping seed, contain a high percentage of water, the tissues of ripe seeds contain only about 15 per cent. of water. During development, the cells of the embryo and of the tissues which accompany it manifest all the properties of actively growing cells. After ripening, it may be said that the seed passes into a condition of " suspended animation " or trance, during which the tissues manifest none of the ordinary activities of life, although there is reason to believe that vital processes continue to take place, albeit so slowly that they can only be recorded by very delicate and careful observations. The cells of a young seed, like those of other actively-growing tissues, are easily injured by extremes of heat and cold, severe drought, or other untoward conditions. In the condition of *dormant life* into which a seed passes when *ripe*, the cells of the embryo can endure severe desiccation, and may be subjected to extremely high or low temperatures without injury.

Vitality of Seeds—The length of time during which the embryo of a dry seed can retain its vitality is determined partly by the nature of the seed itself and partly by the conditions of storage. The seeds of certain plants must be sown as soon as ripe, and unless then provided with the conditions necessary for germination quickly perish. Other seeds refuse to germinate until they have undergone a period of rest. The seeds of a majority of plants, if stored under dry and suitable conditions, retain their vitality for considerable periods, in some cases for many years. Although seeds can be stored for long periods it is not wise for practical purposes to do this, and farmers and gardeners do well when they ask to be supplied with seeds of the preceding season. Owing to vital processes which go on with extreme slowness in the resting seed, changes take place which impair the vitality of the embryo and in time lead to its death. It is for this reason that a high proportion of seeds in a sample of old seed either fail to germinate at all, germinate more slowly than they ought, or give rise to weakly plants incapable of vigorous growth.

For germination purposes, a perfect sample consists of seeds all of which will grow into strong healthy plants. Commercially, it is very difficult to obtain seed of this standard, and the quality of a sample can be judged by making germination tests, and estimating the proportion of seeds which germinate satisfactorily within a given period. The standard thus arrived at is called the " germination capacity " of the sample of seed (experiment 7, p. 158).

The exact nature of the changes undergone by the protoplasm during ripening is not fully understood. It is evidently closely related to the temporary loss of a high proportion of the water contained in active protoplasm and is of the same nature as that suffered by the spores of Bacteria, Fungi, etc., many of which can resist conditions extremely injurious to life and remain in a " resting " or dormant state for long periods.

In certain cases an embryo may develop abnormally without the act of fertilisation having taken place. The production of new individuals in such a manner is sometimes called *parthenogenesis ;* the course of events is both complex and variable and cannot be considered here.

Fruit Formation—The formation of *fruit* by a Flowering Plant is usually incidental to that of *seed*, and in most plants fruits are not formed unless fertilisation has previously taken place and the ovules are developing into seeds. In a few cases, however, fruits are developed whether fertilisation has or has not taken place. This is so, for example, in the Cucumber plant, which forms cucumbers freely with or without seeds. Of a similar nature are certain " seedless " varieties of Banana and Orange.

Advantages of the Seed Habit—A clear understanding of the complex processes which precede and follow fertilisation in Flowering Plants and control the formation of seeds and fruits, helps us to understand many of the difficulties which beset the cultivator of plants. We can understand, for example, why a sharp frost in the spring may cause such widespread damage to the expected fruit crop, if it comes shortly after the flowers of the fruit-trees open. At this stage the stigmas are composed of delicate, succulent tissues, and they are dusted with germinating pollen grains, the tubes of which are very susceptible to injury from cold. If the tissues of the stigma suffer damage, pollen grains cannot germinate upon it; if the pollen grains before or after they germinate are injured, fertilisation cannot take place. When the gardener says his apple and pear trees are safe from damage because the blossom has " set," he means that pollen has germinated safely, and that fertilisation of the egg-cells within the ovules has thereby been effected.

The whole purpose of flowers to the plant has been served when this is ensured, and it is not therefore surprising to find that many flowers fade soon after pollination, and can be preserved from fading for a longer period by protecting the stigmas from pollen.*

We can understand also the immense significance to the plant of the *seed habit*, and the predominance of seed-bearing plants in the vegetation of the world to-day. Plants such as the Ferns are handicapped in competition with Seed Plants by having *two* critical periods during their lives: one at the fertilisation stage of sexual reproduction, and another at the establishment of the young plant at germination. The majority of Ferns are terrestrial plants, but at both these periods they require very moist conditions. At the first, water is required to enable the male gametes,—resembling in their behaviour those of a Brown Seaweed,—to reach the egg-cell and fuse with it; at the second critical period water must be present *as soon as* the embryo plant becomes independent. The young fern plant cannot await conditions favourable to germination and growth, as can the embryo of a seed plant sheltered within the seed. Except under moist conditions of climate, the only Ferns which can compete successfully with Seed Plants are those which possess a really effective means of vegetative reproduction, as does, for instance, the Bracken (*Pteris aquilina*).

Vegetative Reproduction and Sexual Reproduction—Summary —Plants can multiply rapidly and effectively by vegetative means. The production of new individuals in this manner puts less strain upon the plant, and the risks involved by the necessity for pollination and fertilisation are avoided. On the other hand, new individuals have, as a rule, little chance of removal from the neighbourhood of the parent and are liable to suffer from lack of room and overcrowding. In the case of water-plants, these drawbacks are not so

---

* Remarkable examples of such behaviour can be found in some orchids, the flowers of which, if protected from pollination, may remain upon the plant, fresh and unfaded, even for several months.

marked, and reproduction by vegetative means is accordingly very effective in aquatic Flowering Plants, many of which multiply rapidly in this way without recourse to seed formation.

Reproduction by seed puts a severe strain upon the plant at the flowering and fruiting periods. It results in the formation of a larger number of new individuals and, in the case of many plants, there is a better chance that these will be dispersed to some distance from the parent. The most characteristic feature of sexual reproduction is the *dual origin* of each new individual, and the possibilities for the production of new characters, due to increased vigour and new combinations of characters found in the parents, which this affords. These matters will be more fully discussed in Chapter XII.

Practical Work

It is assumed that organs of vegetative reproduction such as rhizomes, corms, tubers, etc. are familiar to the student. The following exercises are intended to supplement the examination and drawing of such specimens.

1. Make "cuttings" of Geranium, Rose, Willow, or other suitable plants. Plant in moist sand, keep moist and warm, and examine once a week to see the formation of a new tissue or "*callus*" over the cut surface. Note later the formation of roots from this callus. Make similar cuttings from twigs of Willow. About two inches from the base of each remove carefully with a sharp knife a ring of bark. Place the ends in water and keep the injured part of the twigs in moist air by covering the vessel with a sheet of cork or other material, which can be perforated to allow the shoots to pass through. Notice the position in which roots form, and make inferences as to the tissues concerned in conduction of food and their position in the plant.

2. Make root-cuttings of Seakale, Dock and Dandelion by cutting the root into pieces of two or three inches in length. Plant in boxes of damp sand, keep moist and examine at intervals. Note the formation of a callus over the cut surface and the origin of new buds and roots from this tissue. Notice specially the position of these with reference to the original root—whether at the upper or lower end only, or indifferently at both ends; test the effect of planting the cuttings in different positions. Make similar cuttings of Carrot or Parsnip and note the result in each case.

3. Determine the effect of pollination as follows. Carefully remove with a pair of forceps the unopened stamens from young flowers of any convenient plant, e.g. Wallflower, Stock, Foxglove, Snapdragon, Evening Primrose, and cover the flowers so treated

with bags of muslin or thin paper to exclude insects. Treat other flowers in a similar way but, before covering, dust the stigmas with pollen from a flower of the same species. After about a week or as soon as the stigmas have withered, remove the bags and record the subsequent formation or not of seeds and fruits in each case. Compare the behaviour of flowers so treated with that of others of the same species treated as follows: (a) stamens removed but flowers not covered or artificially pollinated; (b) stamens not removed but flowers covered with bag. Draw inferences as to the general occurrence of self or cross pollination in the case of each species. If cross pollination is indicated, make observations to try to determine the agent concerned in the transference of the pollen.

4. Examine the pollen of various plants by mounting a little in water or weak sugar solution and examining microscopically. Note differences in size and shape of grains and in the structure of the outer wall. The following notes may be found useful. *Very large grains* : Mallow, Hollyhock. *Walls sculptured with spines, warts, etc.* Hollyhock, *Dahlia*. *Thin walls* : Pea, *Primula*. *Grains attached to one another by mucilaginous threads* : Fuchsia, Evening Primrose, *Rhododendron*. *Grains adhering together in groups of four (tetrads)* : Heath, Ling, and many plants of the Heath family, Bullrush (*Typha sp.*), *Acacia*. *Grains with thin places or pits of remarkable structure* : Cucumber, Marrow, White Bryony. *Pollen adhering to form a solid body or pollinium* : Orchids (with rare exceptions), *Asclepias*.

5. Germinate pollen grains artificially; examine microscopically and note the formation of pollen-tubes and the movement of the contents of the pollen grain into them.

Pollen can be germinated by placing a little sugar solution in a small watch-glass, covering with another watch-glass to exclude dust, and standing in a warm room; after some hours, or on the following day, examine drops of liquid from the watch-glass for germinated grains. A better method is to make " *hanging drop* " *cultures*. Prepare some *perfectly clean* glass slides and square cover-slips. Cut a few strips of blotting paper to fit each slide and cut a hole in the centre of the " pad " so formed with either a sharp chisel or a cork-borer. Take a clean cover-glass, lightly smear the *edges* of it with vaselin; place with a glass rod in the middle of the vaselined side a small drop of sugar solution. Transfer with a needle a *little* pollen of the kind desired from a freshly-opened anther. Moisten the blotting-paper " pad " with clean, freshly-boiled water and quickly invert the cover-slip over the hole, so that the drop is suspended in a moist chamber formed by the blotting-paper. Pollen grains do not germinate satisfactorily in water, since they usually contain much sugar or other osmotic substances in the cell-sap. If placed in water they take up water, swell, and frequently burst. The strength of sugar solution required varies with the grains used and must be determined by experiment. The pollen of species of *Sedum* and

Echeveria, plants often grown in the rock garden and for carpet-bedding, is very suitable for class-work. It germinates well in a sugar solution of about 15 per cent. strength and with unusual rapidity. It is advisable when making " hanging drops " of this kind to keep the conditions as *aseptic* as possible. Use freshly-boiled water and sugar solution and pass slides, cover-slips, needles, etc. through the flame of a spirit lamp or a bunsen burner before use. After microscopic examination, the slides should be placed in a covered glass vessel or other suitable receptacle on a piece of moistened blotting-paper and examined at intervals.

" Hanging drops " of the spores of various Fungi, etc. can be made in a similar way and the germination observed.

6. Interesting information can be acquired by making named collections of seeds of common plants. It is a useful exercise for the student to separate the seeds of small " mixed lots " of such seeds and identify them by comparison with those in the collection. The seeds should be examined with a good lens, drawn, and any peculiarities of shape, sculpturing of seed-coat, etc. noted.

7. Determine the germination capacity of various samples of seed by germinating representative lots of ten on blotting-paper in a germinator and estimating the percentage germination from the behaviour of e.g. five such lots (see p. 67).

8. The common Chickweed is an *ephemeral,* i.e. it produces several generations in the course of a single growing season. Collect seeds from successive generations, especially those early and late in the season, and compare their behaviour when germinated, (*a*) immediately after collection, (*b*) six months later.

CHAPTER XI

THE OUTLINES OF CLASSIFICATION

THE BASIS OF CLASSIFICATION. ARTIFICIAL AND NATURAL SYSTEMS.
THE LINNÆAN SPECIES; ELEMENTARY SPECIES; VARIETIES. THE
OUTLINES OF CLASSIFICATION OF LIVING PLANTS. TAXONOMY AND
HEREDITY

The Basis of Classification—When we look around at the immense variety of living things in the world our first impression is apt to be one of confusion and doubt as to the possibility of extracting order from such diversity. A small amount of observation, however, convinces us that living things can be separated into groups, the members of which have more in common with one another than they have with the members of any other group. Further study of any one of these groups shows that it is possible to subdivide it further along similar lines, and proceeding in this way we can evolve a *classification* of living things, the simplest unit of which is called a *species*.

Thus, organisms as a whole fall naturally into two great groups, the Animal Kingdom and the Vegetable Kingdom, the typical members of which are strikingly dissimilar in many respects. Now, it is generally believed by biologists that all plants and animals have descended from a common ancestral stock, which in turn can be traced back to the earliest and most primitive forms of life, as to whose manner of origin we have no knowledge whatsoever. Facts such as the following confirm this view.

Although the higher members of the two kingdoms are so unlike, a study of plant nutrition will have convinced us that plants and animals have much in common. This impression is strengthened by a study of the simplest plants and animals, some of which are claimed equally by zoologists and by botanists. It often becomes necessary, indeed, when dealing with some of the simple forms of life to make arbitrary distinctions, and the position assigned to such and such a group of lowly organisms may depend upon the importance attached to some such distinction, or upon a general agreement among biologists to accept it as a matter of convenience.

The contrast between animals and plants may be pictured as due to the operation of divergent tendencies which appeared very early in the evolution of living things,—tendencies which have operated in such a way as to widen the gulf between the more advanced members, so that the typical plant and the typical animal are very different indeed, whereas if the line of descent is traced backwards in either group, the width of the gulf tends to diminish.

Artificial and Natural Systems of Classification—Attempts to classify plants began with the Herbals of the sixteenth century, but the older systems of classification were based entirely upon superficial resemblances and were, in consequence, quite artificial. They served the purpose of their designers, viz. that of classifying living organisms so that they could be named, described and recognised; the merits of a system were judged by its convenience in these respects.

Recognition of the principle of evolution profoundly modified the purpose of systems of classification. The object of all modern systems is to group organisms in such a way as to indicate their real relationships or affinities.

The purpose of a complete classification of plants is not merely, therefore, to provide a mechanical arrangement convenient for reference like a card-index, but is to construct a genealogical tree of the Vegetable Kingdom. Even modern classifications are far from perfect when criticised from this point of view and the difficulties which present themselves are very great. The construction of a genealogical tree, complete in all its details, demands not only an exhaustive knowledge of the plants now in existence, but also a complete historical record of those of past ages. The groups of plants now in existence can be represented as the ultimate branches of such a tree, although it is not always possible to show the relations of these branches with one another, or to indicate how they are connected with the common stock or trunk.

There is inevitably an element of artificiality even in a Natural System of Classification, since one group may merge into another and the limits of closely related groups are not always sharply defined.

In systematic botany the modern period may be said to begin with Linnæus, the famous Swedish naturalist of the eighteenth century. The great contribution made by Linnæus to Botany was not the artificial system of classification known as the *Linnæan system*, now long superseded, but was due to his wonderful gift for describing and classifying plants, or indeed any other objects which engaged his attention. Linnæus recognised that his method of classification was artificial, and declared that the chief task of botanists was to discover a natural system. His own work was to bring order out of chaos, and to name accurately and describe as many species as possible. Although Linnæus did not invent what is known as the *binary* method of naming plants and animals, he first applied it to practical use, and introduced order and clearness into the art of describing plants, an essential step towards classifying them correctly. In the binary system which is now universally used, each plant or animal is designated by two names; the first, the *generic* name, is that of the *genus ;* the second, the *specific* name, is that of the *species*. Thus, *Ranunculus repens* is the name of a Buttercup belonging to the genus *Ranunculus* which includes a number of species. The general characters are those of the genus *Ranunculus ;* the specific name " *repens* " has reference to the creeping habit which distinguishes this species from several others to which it bears a close resemblance.

It is important at this stage to have a clear idea as to what is meant by a *species*, although it is not easy to frame a definition which will meet with general acceptance. Linnæus had a perfectly clear idea of what he meant by a species, which he defined as a group of plants (or animals) which owed its origin to a special act of creation. He assumed that each of the vast number of kinds of living organisms in the world, sharply distinguished from others by specific characters, had come into existence " ready-made," so to speak, and continued to exist in its original form.

The systematic botanist still makes use of the Linnæan species as a basal unit of classification, but it is generally recognised that such species are often connected by a number of intermediate forms or varieties which may make it extremely difficult to distinguish them sharply from one another. In

II

other cases there are no such intermediate forms and the limits of a species are readily recognised.

There are cases in which several distinct forms, at one time included within the limits of a Linnæan species, are now regarded by some systematists as good and distinct species; in other cases such forms are called " elementary species " or " petites espèces," or " Jordan's species," after a Frenchman of that name who studied the matter experimentally by cultivating the forms included within certain Linnæan species and observing that the differentiating characters remained constant in successive generations. In yet other cases, the different forms are called *varieties* of the original species and are designated by varietal names placed after that of the species.

The Outlines of Classification of Living Plants—The earlier botanists who attempted to classify plants noticed that many bore the organs we call *flowers*, while others did not. They classified plants, accordingly, into *Flowering Plants* and *Flowerless Plants*, or *Phanerogams* and *Cryptogams*. As knowledge grew, it was recognised that many so-called Cryptogams produce organs comparable with flowers, and that the real distinction lies in the production of *seeds*. The Phanerogams, therefore, are now known as seed-bearing plants or *Spermaphyta;* the non-seed-bearing plants include the Cryptogams of the earlier botanists, and also certain groups of simple organisms unrecognised in earlier times. The Cryptogams are subdivided into the three following groups.

1. *THALLOPHYTA*—Includes *Bacteria, Fungi, Algæ* and *Lichenes*, as well as certain lowly groups on the border-line between the Animal and Vegetable Kingdoms. The members of this group have a *thallus*—i.e. a vegetative body not differentiated into root, stem and leaf.

2. *BRYOPHYTA*—Includes the Mosses (*Musci*) and the Liverworts (*Hepaticæ*). The simpler members of the latter group are thalloid in structure. The Mosses are leafy plants, the most advanced of which possess a rudimentary conducting system. Both Mosses and Liverworts reproduce sexually and show a well-marked alternation of generations. The zygote develops into a sporogonium which remains attached to the sexual plant until the spores are shed. Each spore can give rise to a new Moss or Liverwort plant, and so start the cycle afresh.

3. *PTERIDOPHYTA*—Sometimes called the "Vascular Cryptogams." Includes the Ferns (*Filicales*), the Horsetails (*Equisetales*) and the Club Mosses (*Lycopodiales*). All members of the group are vascular plants showing a well-marked alternation of generations. The fern plant or its equivalent in the other groups produces spores, each of which can give rise to an independent prothallus or sexual generation bearing gametes. The zygote resulting from fertilisation develops into a new fern plant or sporophyte.

SPERMAPHYTA—The seed-bearing plants or Spermaphyta comprise the vast majority of vascular plants, and represent the highest stage of structural development in the Vegetable Kingdom. In all systems of classification Spermaphyta is subdivided into *Gymnosperms* and *Angiosperms*, distinguished by the manner in which the seeds are borne.

Gymnospermæ—Includes three groups: *Cycadales, Coniferales* (including *Gingkoales*) and *Gnetales*. The affinities of these sub-groups are not apparent and their association in a single group is open to criticism. The majority of living Gymnosperms belong to *Coniferales*, the members of which are evergreen trees of characteristic habit. The seeds are borne either on small branchlets, as in the Yew, or at the base of woody scales which are grouped together to form cones, as in the Pines, Firs, Larches etc.

Angiospermæ—Includes the majority of seed-plants, which are the most abundant and successful plants in the vegetation of the world. The members of this group are trees, shrubs or herbs; they are distinguished from the Gymnosperms by the fact that the seeds are borne within a closed organ, the ovary, formed by the carpels of the flower.

In all systems of classification the Angiosperms are divided into two great classes, *Monocotyledons* and *Dicotyledons*, the members of which are distinguished by certain constant structural differences. Monocotyledons and Dicotyledons are further subdivided into successively smaller groups, until the unit of classification known as a *Family* or *Natural Order* is reached. A Natural Order is defined as " a group of *genera* which resemble each other more than they do other genera."

The genera included in a Natural Order each comprise one or more *species*. It is convenient to the systematic botanist to adopt the species as a kind of basal unit for purposes of classification. Attention has already been drawn to the difficulty of defining the term " species " as used in modern botany. Most Linnæan species, when carefully observed, are found to include forms or " little species " as they are called and also varieties, the latter distinguished by minor differences of habit or structure. It is evident, on reflection, that this is to be expected, since the great natural laws which control and direct evolution must be constantly operating upon the plants now in existence.

Systematic Botany or *Taxonomy* is concerned with the accurate description and classification of individual plants according to their natural affinities. The manner of origin of the different forms and the nature of the laws governing their appearance is the concern of the student of evolution and heredity. A knowledge of the manner of operation of these laws is of profound theoretical interest and the practical application of it has proved of great value to breeders of plants and animals.

Practical Work

The student should endeavour to acquire a working knowledge of plant classification by means of field excursions, using a flora to identify species and noting carefully the habitat and distribution in each case. Examination of plants in the field should be supplemented by careful study and dissection in the laboratory.

As work proceeds it will be found useful to select some convenient genus for special study and make an herbarium collection of any different forms represented among its species.

CHAPTER XII

EVOLUTION, VARIATION AND HEREDITY

EVOLUTION AND THE "ORIGIN OF SPECIES." VARIATION: NATURE AND ORIGIN OF VARIATIONS. MUTATIONS. HEREDITY: SEXUAL REPRODUCTION IN RELATION TO HEREDITY. MENDELISM. THE CHROMOSOMES IN HEREDITY: RELATION BETWEEN THE CHROMOSOMES AND THE MENDELIAN FACTORS.

Evolution—The problem of accounting for the immense number of different forms of life in the world has aroused the interest of thinkers in all ages. The view that the diversity has come about by a process of evolution was propounded by Aristotle in the fourth century B.C. The group of great thinkers and naturalists of the nineteenth century, who propounded the doctrine of organic evolution in its modern form, secured its general acceptance in place of the doctrine of special creation already referred to in connection with the work of Linnæus (p. 161).

On the theory of evolution, how is it possible to account for the immense diversity of habit, structure and function among living things ?

There are various ways in which the problem may be attacked: by speculation, by observation, or by experiment. It is clear that speculative theories are valuable in proportion as they are based upon accurate observations, while confirmation by experiment is the most convincing proof that can be offered.

One of the earliest attempts to explain the problem of the " origin of species " was made by Lamarck in the early years of the nineteenth century. Lamarck contended that the organs of animals and plants are modified by use (in the case of plants, by environment), that such modifications are inherited by the offspring, and thus, in the course of generations, give rise to new species. There is abundant evidence to establish the truth of the first part of this hypothesis;

no one who has observed living organisms growing under different conditions can doubt that their surroundings exercise a profound effect upon the manner of their growth and development.

It is not possible to deal here with the vexed question of the " inheritance of acquired characters," but it may be said that, although there is a popular prejudice in favour of the view that such characters are handed on from parent to offspring, there is no generally accepted evidence that such inheritance actually takes place.

After the time of Lamarck, biologists had to wait another half-century for a theory of the " origin of species " based upon accurate observations on living organisms. In 1859 Charles Darwin published the " Origin of Species," and his views, in so far as they are opposed to those of Lamarck, may be summarised as follows.

(1) *Small variations occur* among the offspring of every individual.

(2) There is, in the case of all species living in a state of nature, an *over-production of offspring*, i.e. there are born into the world many more individuals than there is room for.

(3) There is, in consequence, a *struggle for existence*, in which those individuals which have varied in a way serving to fit them better for their surroundings will, on the whole, have most chance of survival. These will be the parents of the next generation, to the members of which will be handed on the characters that determined the survival of their parents.

The two principles, " the struggle for existence," and " the survival of the fittest " due to the operation of *Natural Selection*, exercised a profound effect upon the study of evolution and the origin of species.

It may be noted, in comparing the views of Darwin with those of Lamarck, that the former does not attempt to explain how the variations upon which Natural Selection operates arise in the first instance; he simply states, supporting his statement by a vast array of facts, that small, continuous variations do occur in nature.

Variation—Since the time of Darwin and the general acceptance of the theory of Natural Selection, interest has shifted to a consideration of the nature and origin of the variations upon which Natural Selection acts, and the manner in which characters are handed on from one generation to another.

An examination of any population soon shows us that the variations which occur can be classified under two heads, *continuous* and *discontinuous* variations. By a continuous variation is meant one in which individuals showing every degree of development of the character in question can be found between the two extremes: a discontinuous variation, on the other hand, is not connected with the type by a series of intermediate forms but is sharply marked off from it.

Different species of animals and plants show very unlike degrees of variability; in some the characters are very constant; in others there is a marked tendency to vary,—closely allied species frequently behaving quite differently in this respect. Thus, as regards the relative amounts of black and red on its wing cases, the insect called the " seven-spot " Ladybird shows remarkable constancy, while the species known as the " two-spot " Ladybird shows an extraordinarily wide range of variability,—insects being found which show all stages of variation in the wing-cases between two black spots on a red ground and two red spots on a black ground ! As a frequently encountered example of discontinuous variation in plants, the Milkwort (*Polygala vulgaris*), which occurs commonly in forms with blue, red, or white flowers, may be cited.

Let us now briefly consider what is known as to the nature of variations and the ways in which they arise.

Nature and Origin of Variations—The modern views regarding variation in plants and animals may be outlined briefly as follows:—

(1) There are individual variations traceable to the effects of the environment, such, for instance, as the differences observable between plants of the same variety when grown in sunny and shady situations. According to the Lamarckian theory, such variations are inherited and can thus give rise to new species. There is, at present, no entirely satisfactory evidence that modifications of the body of an individual acquired in this way can be transmitted to the sex-cells and, through them, handed on to the next generation.

(2) There occur small individual variations which cannot

be referred to effects of the environment, and which tend to be continuous. It is chiefly upon these small, ordinarily unnoticed variations of a continuous type that the law of Natural Selection, according to Darwin, operates.

(3) Varieties differing sharply from the type, and not connected with it by a series of intermediate forms, are also met with. These " discontinuous " variations may owe their existence to several causes. Firstly, they may be the isolated remnants of a continuous series, the intermediate members of which are absent for some reason. Secondly, they may arise as a result of hybridisation, concerning which we shall have more to say later. Thirdly, they may arise suddenly and sporadically as isolated " sports," entirely unconnected with the type by intermediate forms. The existence of " sports " was clearly recognised by Darwin, who regarded them as playing a minor part in the production of new species.

Interest in the last class of discontinuous variations was revived by the investigations of the Dutch botanist de Vries. As a result of observations on wild and cultivated plants of the Evening Primrose (*Œnothera Lamarckiana*), he found that a small percentage of the seedlings (rather over 1 per cent.) differed from the majority, and that these exceptional plants passed on their special characteristics to their offspring. De Vries does not regard the Evening Primrose as a special case, but considers that any species of plant, if grown in large numbers, will give rise to a small proportion of such *mutations*, the actual percentage depending upon the species of plant, the conditions of cultivation, and so on. According to this view, new forms, differing in many particulars from the type, originate suddenly and spontaneously as seed-variations and are capable of handing on their peculiarities to their offspring. According to de Vries, Natural Selection is not concerned with the appearance of these characters as a result of the slow accumulation of minute differences; it is only after their spontaneous appearance that Natural Selection becomes operative. " Natural Selection may explain the survival of the fittest, but it cannot explain the arrival of the fittest."

It is often difficult to account for the appearance of a new species as a consequence of the accumulation of small, continuous variations because, in many cases, the variation may be of such nature or so insignificant that it would seem to possess doubtful " survival value." If, however, as de Vries suggests, new forms, differing widely from the type, can appear " at a jump," Natural Selection is provided with a much wider range of material upon which to work, and a reasonable explanation is forthcoming to account for the occurrence of varieties possessing characteristics of no obvious value. To what extent the sudden appearance of new forms is to be attributed to the effects of mutation or is the result of hybridisation is a vexed question which we have not the time to consider, but which will be appreciated more fully towards the end of this chapter.

In conclusion, it must be pointed out that, besides these seed variations, vegetative or bud variations occur, which may likewise form the starting point of a new race. Thus, a branch bearing flowers of a different colour or shape may suddenly appear on a plant. The name " sport " is used popularly to describe all cases of apparently spontaneous variations, whether they arise as seed or vegetative variations. It is only correct to describe a vegetative variation as a mutation when it can be transmitted by means of seed.

We come now to a third aspect of variation, namely, the manner in which the characteristics, originating as the result of some kind of variation, are inherited.

Heredity—Heredity is the branch of science concerned with the study of *inheritance*, or the manner in which characters are handed on from parents to offspring.

It is a commonplace that offspring of the same parents resemble one another and their parents more than they do unrelated individuals. It is well known, also, that members of the same family are not all exactly alike and that they show individual peculiarities distinct from those of the parents. In passing from one generation to another, therefore, we encounter the phenomena of *inheritance* and *variation*. Variation amongst the members of one family can sometimes be accounted for by assuming that differences occur in the degree in which the parental characters are inherited, resulting in different combinations of such characters in the offspring: also, what may be described as *ancestral characters*, not evident in the parents although " latent," may reappear in certain of the offspring.

The study of heredity, variation and everything relating to this aspect of living organisms is known as the science of *Genetics,*—a recent branch of Natural Science which, apart from its absorbing theoretical interest, has already yielded results of great practical value.

The material used in the study of genetics includes the whole realm of organic nature: in the limited space at our disposal we can only attempt to indicate a few of the more salient points of interest concerning inheritance.

There are two ways in which we may attempt to discover the laws governing inheritance. As a result of observations carried out upon a population as a whole we may obtain an idea as to how a particular character is inherited *on the average*, although we shall not be able to predict with certainty its behaviour in the case of any particular individual. The alternative to this *observational* method, in which the results are subsequently treated statistically, is the *experimental* method, in which the behaviour of the character in question is followed in particular individuals from one generation to another. The latter method is more lengthy and laborious, since it entails the carrying out of breeding experiments for several generations; furthermore it is not completely applicable to all cases (the inheritance of human characteristics, for example), but it has the great advantage that it gives results which can be applied to individual cases; i.e. if the nature of the parents is known, it is possible to predict not only the *types* of offspring but the relative numbers in which these will appear. The two methods are not, of course, mutually exclusive, for results obtained by one method can often be usefully supplemented by those obtained by the other. In what follows we shall consider the experimental method, since not only are the results yielded more precise, but they can be correlated with the facts already learned concerning sexual reproduction.

Sexual Reproduction in Relation to Heredity—Every plant (or animal) formed as a result of sexual reproduction begins its existence as a fertilised egg-cell or zygote, formed by the union of two sex-cells or gametes. It is clear, therefore, that all the characteristics of the adult must be present in the zygote and become evident as development proceeds.

In Flowering Plants, the process which culminates in fertilisation is initiated when a pollen grain reaches the stigma of the same or another flower. The plant producing the pollen grain is called the male parent; that bearing the stigma and ovules is called the female parent. When the pollen of one plant is used on the stigma of another, the operation is known as *crossing* one plant with the other. Many species of Flowering Plants, however, bear both stamens and ovules on the same individual (i.e. they are *hermaphrodite*), and the transference of pollen to the stigma of the same flower or to a different flower of the same plant is known as *self-pollinating* or *selfing** the flower.

In order to be able to refer shortly to the different generations in a breeding experiment it is usual to call the first generation the *F1 generation* (first filial generation), while the family to which this gives rise in its turn is called the *F2 generation*, the next the *F3 generation*, and so on.

Mendelism—It is convenient to commence our study of the problems of inheritance in plants with a description of an actual breeding experiment. The discovery of certain fundamental laws which govern heredity resulted from experiments on the mode of inheritance of *tallness* and *shortness* of the shoot in ordinary garden peas, and we cannot do better than make use of the same material for our inquiry.

If we grow a plant of a " pure strain " of tall-growing peas and self the flowers, the seeds, when sown, will all develop into tall-growing plants,—a result which will be obtained consistently for as many generations as the experiment is carried on. This, indeed, is what is meant by saying that the strain is " pure " for the character of " tallness " (see p. 174). Treated in a similar manner a short-growing or dwarf variety of pea gives rise only to " dwarfs."

Suppose, now, that we cross a tall pea with a short pea, either transferring the pollen from a flower on the tall plant to the stigma of a flower borne by a dwarf plant, or *vice versa*. The seeds produced from such a cross give rise to tall plants indistinguishable from the tall parent, although they behave very differently when used for subsequent breeding experiments.

A second generation (F2) can be obtained from the first generation (F1) in any of the three following ways, each of which yields a different result as regards the character of the offspring :—

* In genetics this term is often extended to include the crossing of two individuals exactly alike.

(a) An F1 plant is selfed (i.e. crossed with itself). This gives rise to a subsequent (F2) generation containing *both tall and short plants in the approximate ratio of* 3 : 1.

(b) An F1 plant is crossed with a plant of the original tall strain. This gives rise to an F2 generation, *all the members of which are tall.*

(c) An F1 plant is crossed with a plant of the original short strain. This gives rise to an F2 generation containing *both tall and short plants in approximately equal numbers.*

It is evident from these results that the tall plants obtained in F1, although apparently similar to the original tall parent, are in reality of a different constitution, seeing that, when selfed, they give a proportion of dwarf plants. These two kinds of tall plants are described respectively as *homozygous* or " *pure* " for tallness and *heterozygous* for tallness. The significance of these terms will be evident later. Further, since the shortness does not show in the heterozygous tall plants, although subsequent breeding shows it to be present and capable of being transmitted to the offspring, tallness is said to be *dominant* to shortness (or conversely, shortness is *recessive* to tallness). It can readily be shown by breeding experiments with the F2 generation produced as described in (a) and (b), that the tall plants, although apparently all alike, are really of two kinds, viz. homozygous and heterozygous for tallness like the F1 plants. Since, when selfed, the homozygous tall plants give rise to tall plants only, while the heterozygous tall plants give rise to both tall and short plants, it is possible to distinguish readily between the two types of tall plants by breeding. Ascertained in this way, the facts may be expressed diagrammatically as follows:—

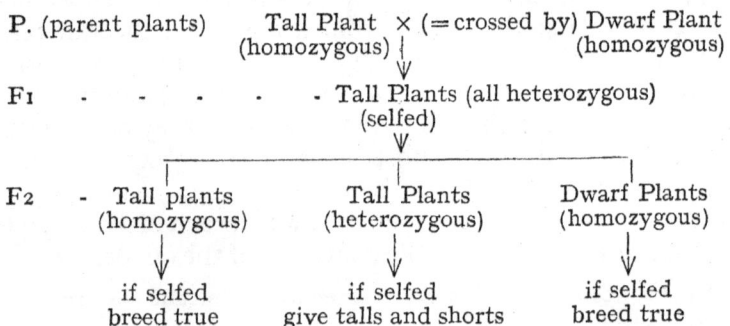

P. (parent plants)   Tall Plant   × (= crossed by) Dwarf Plant
                    (homozygous) | (homozygous)
                              V

F1   -   -   ·   ·   - Tall Plants (all heterozygous)
                            (selfed)
                            V

F2 - Tall plants	Tall Plants	Dwarf Plants
(homozygous)	(heterozygous)	(homozygous)
↓	↓	↓
V	V	V
if selfed	if selfed	if selfed
breed true	give talls and shorts	breed true

It is evident from these experiments that the character of tallness is inherited in a regular way. The discovery by experimental methods that the transmission of characters from one generation to another takes place according to definite rules was made by an Austrian monk, Gregor Mendel, who was Abbot of the Königskloster in Brunn from 1868 until his death in 1884. His experiments were ignored at the time, and it was only comparatively recently that their full significance was appreciated. The branch of Genetics known as *Mendelism* is founded upon the breeding experiments carried out in Mendel's garden, and marks an epoch in the study of the problems of heredity.

The great importance of Mendel's work lay in the fact that it provided for the first time an experimental method for dealing with problems of heredity. This, forming the starting-point for much subsequent research, has profoundly modified the study of inheritance.

As a result of experiments on the lines described, Mendel suggested a scheme to account for the way in which characters are inherited. According to this, an individual is to be regarded as possessing a great number of separate and distinct unit characters which, taken all together, give to it the features by which it is known. Each of these unit characters is considered to be the expression of a definite " something " in the plant called a *factor*. Thus, the characteristics of an individual may be regarded as the outcome of the particular factors it possesses, the factors being contributed to the zygote by the gametes. For example, tall stature, coloured flowers, a hairy stem, dwarfness, etc. are due to the presence of the factors for tallness, colour, dwarfness, and so on. It must be clearly understood that the possession of a particular factor is held to represent a *capacity to develop* a particular character under appropriate conditions and is not the character itself; it may happen that conditions prevent a factor which is present from exercising its capacity and showing itself as a definite character in the plant.

Now, certain pairs of characters, such as tallness and dwarfness, colour and absence of colour in the flowers, and so on, are mutually exclusive and stand to one another in a special relation. Mendel called such alternative factors *allelomorphs*, and believed that while a gamete could carry either of two such factors, it could not carry both at the same time. To take an example from those already cited, the factors for tallness and dwarfness form an allelomorphic pair, i.e. a gamete can carry the factor for tallness or the factor for dwarfness, but not the two factors simultaneously.

Let us consider now the behaviour in inheritance of a pair of characters such as tallness and dwarfness, in the light of what we have learnt regarding sexual reproduction. The gametes, whether male or female, produced by a pure tall plant all contain the factor for tallness, which we may represent by " T." When such a plant is selfed, a male gamete containing T meets a female gamete containing T and a zygote results containing two doses of T (i.e. TT): this develops into a mature plant every cell of which contains TT. When the time arrives for this plant to form sex-cells, the two doses of T separate, according to Mendel, with the result that each male gamete and each female gamete contains only one dose of T. To summarise briefly; a pure tall plant contains two doses of T; there is a complete separation of these doses previous to the formation of gametes, with the result that *all the gametes are alike* in respect to this factor.

We can define now more exactly what is meant by a plant being " pure " in regard to any character. It is that *all* the gametes produced by the plant contain *one* dose of the factor for the character in question, and consequently each of the offspring produced by selfing such a plant will contain *two* doses of the factor and will be identical with the parents so far as this character is concerned. If we represent the factor for dwarfness by *t*, then, in a similar way, the pure dwarf plant contains two doses of *t* (i.e. *tt*) and each gamete produced by it contains one dose of *t*.

Now consider what occurs when a pure tall plant is crossed with a pure dwarf plant. The female gamete from the tall plant contains T and the male gamete from the dwarf plant contains *t* (or *vice versa*): the resulting zygote therefore contains one dose of T and one of *t*, i.e. has the constitution T*t*. Into what sort of plant will this zygote grow? How will the opposite tendencies represented by T and *t* counteract one another? And how will the factors be distributed to the gametes?

In the first place, we have learnt from our experiments that this plant will be indistinguishable in appearance from the tall parent,—in other words, the factor T is completely *dominant* over the factor *t*. This type of plant, with one dose of each factor, is called a *heterozygote* to distinguish it from a *homozygote*, in which only one of the two allelomorphic characters is represented, although that is present in two doses.

In the second place, at the formation of gametes the factors separate as before, so that each gamete contains one dose of

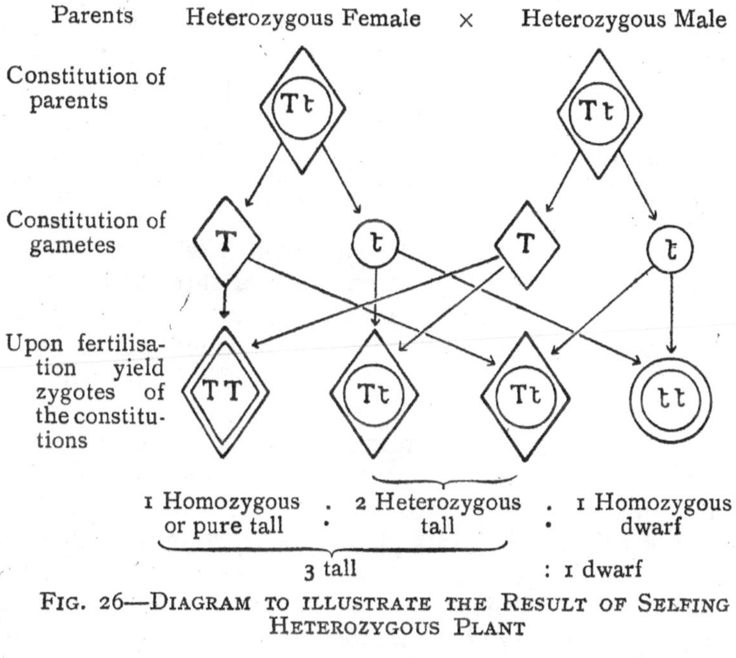

FIG. 26—DIAGRAM TO ILLUSTRATE THE RESULT OF SELFING A HETEROZYGOUS PLANT

one factor only. Consequently, *the gametes produced by a heterozygote are not all alike;* in the present example, half contain T and half *t*.

When a heterozygous plant is selfed, three types of zygote result, thus: (1) a gamete containing T may meet another containing T, yielding a zygote TT; (2) a gamete containing T may meet another containing *t*, giving rise to a zygote T*t*; (3) a gamete containing *t* may meet a similar gamete, resulting in a zygote *tt*. The way in which these matings occur is illustrated in the diagram (fig. 26), on the supposition that one kind of mating is as likely to occur as any other. The relative proportions in which these three types of plant occur is evidently calculable by means of the mathematical laws which govern "chance." Non-mathematical readers may find an experimental illustration helpful. Suppose that equal numbers of black balls and of white balls are contained in a bag, and we draw them two at a time at random. On an average, the numerical proportions of pairs of black balls, to pairs consisting of a black ball and a white ball, to pairs of white balls will be as 1 : 2 : 1 respectively.

It is clear from the diagram that three types of plants are produced in the first generation, the relative numerical proportions being, on the average, 1 pure tall (TT) : 2 hetero-zygous plants (Tt) : 1 pure dwarf (tt). Upon selfing, the pure tall and pure dwarf plants give plants like themselves, while the heterozygous plants will again split into three types in the ratio of 1 : 2 : 1.

In the present instance, the apparent result of selfing is to produce offspring in the proportion of three tall plants to every dwarf, owing to the fact that tallness is completely dominant to dwarfness and the heterozygous plants are only to be distinguished from the pure talls on subsequent breeding.*

In the case of many allelomorphic characters there is not complete dominance of one factor over the other, and the characters of the heterozygotes may be intermediate between those of the two homozygous plants. For example, when a race of plants with coloured flowers is crossed with a strain having white flowers, the flowers of the heterozygous plants, possessing one dose only of the factor for colour, are frequently of a pale shade easily distinguishable from the full colour of the pure coloured strain.

In a few cases the heterozygote is unlike either of the races from which it is derived. The best known case is that of the Blue Andalusian Fowl, a domestic fowl with blue-grey plumage. When two "blue" fowls are mated together, the offspring consist, on the average, of "blue"

* See also Plate III., which illustrates the inheritance of long and short "awns" in "bearded" and "beardless" varieties of Wheat.

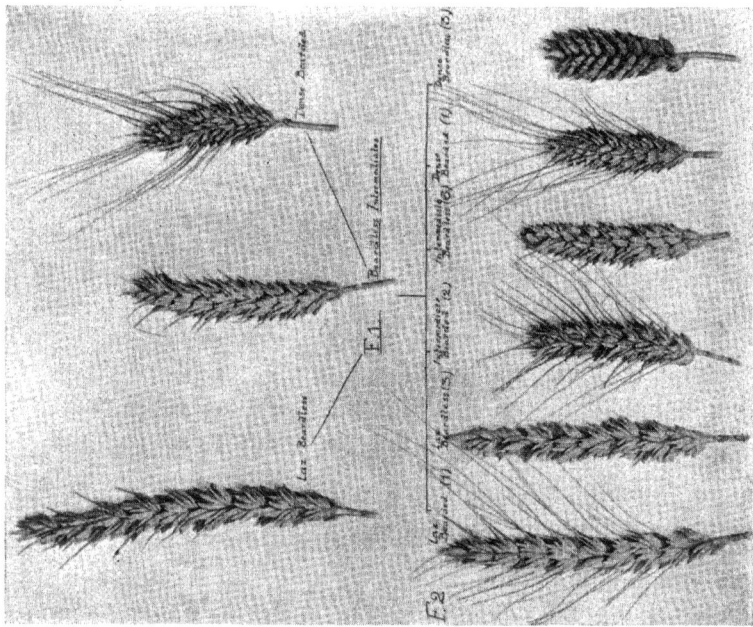

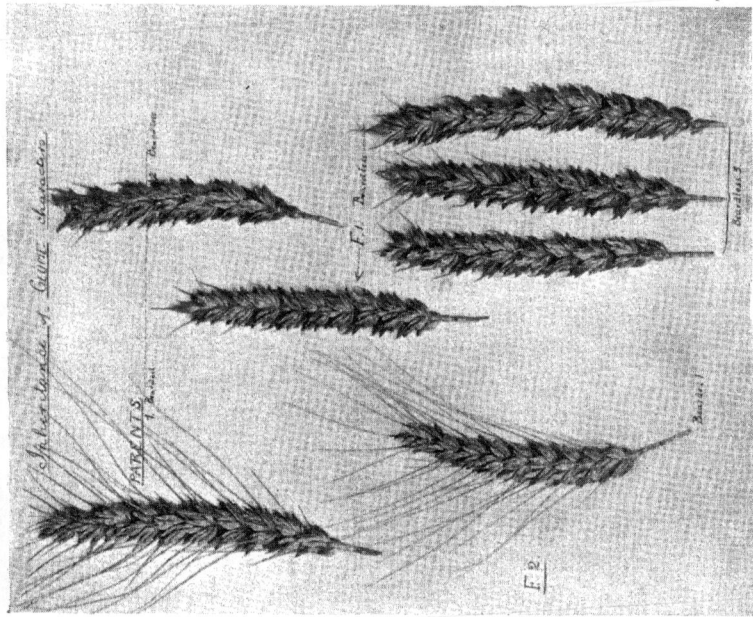

MENDELIAN INHERITANCE IN WHEAT

A, illustrates the behaviour of a single pair of characters—beardless and bearded—the former being completely dominant over the latter.
B, illustrates the behaviour of two pairs of characters – beardless and bearded, lax and dense. Plants heterozygous in respect to the latter pair of characters are intermediate between the two parents in denseness of spike.

birds, black birds and white birds with black flecks on some of the feathers in the proportions of 2 : 1 : 1. Both the latter types breed true when mated with their like; white and black birds mated together give rise to offspring all of which are " blue." It is evident from these results that the Blue Andalusian Fowl is a heterozygote and that all efforts on the part of poultry fanciers to produce a pure breed of it must be so much waste of time. The result of mating together two " blue " birds is analysed graphically in the diagram below, in which the factor for black feathers is represented by " B " and that for white feathers by " b."

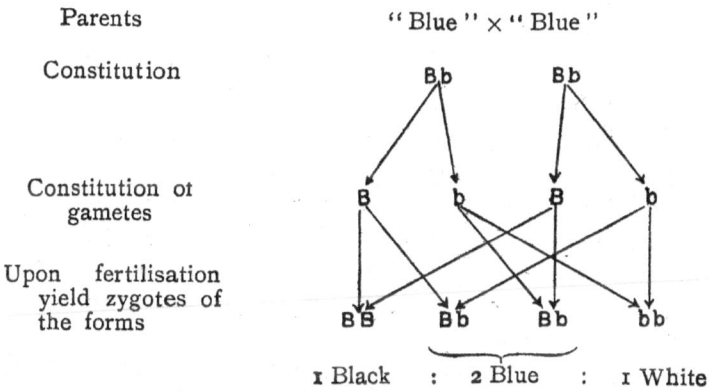

Parents	"Blue " × " Blue "
Constitution	B,b B b
Constitution of gametes	B b B b
Upon fertilisation yield zygotes of the forms	BB Bb Bb bb

1 Black : 2 Blue : 1 White

FIG. 27—DIAGRAM TO ILLUSTRATE GENETICS OF BLUE ANDALUSIAN FOWL

It is, perhaps, hardly necessary to point out that the degree of dominance of one factor over another, although it may determine the appearance of the heterozygote and render this difficult to distinguish from the dominant homozygote, does not affect the manner of inheritance of a character.

The results achieved by Mendel and extended by subsequent workers, not only provide an explanation of the mode of inheritance of pairs of allelomorphic characters, but provide the breeder with a practical method of reaching the result he aims at, and may save much time and labour otherwise devoted to the pursuit of the unattainable, e.g. the production of a pure-breeding heterozygous race, like the Blue Andalusian Fowl.

12

The Mendelian scheme of inheritance is based on the following suppositions.

(1) Every individual is built up of a number of " unit characters," each of which is the expression of the presence of a definite " factor " for that character.

(2) The unit characters can be grouped into contrasted pairs (" allelomorphic pairs "). In respect to any one pair there are three possible types of individuals: (a) those containing two doses of one member of the pair, (b) those containing two doses of the other member of the pair, (c) those containing one dose of each of the pair. Thus, in all cases, the individual possesses a dual structure.

(3) Previous to the formation of gametes, there is a complete separation or *segregation* of the factors making up each pair, thus giving rise to gametes all alike in cases (a) and (b) above, and to gametes of two kinds in equal numbers in the case of (c). Plants of the (a) or (b) type are said to be *pure* or *homozygous* in respect to the character concerned, those of the (c) type are said to be heterozygous.

(4) The rules for determining the number of types and the relative frequency of their occurrence among the offspring may be expressed as follows:—(a) Write down the possible kinds of gametes produced by each parent. (b) Write down all the possible zygotes which can result from their fusion, on the supposition that every type of male gamete meets every type of female gamete once. (c) Group together zygotes which contain similar factors.

The essence of the Mendelian method may be summed up under the following heads:—

(1) One or a few characters only are investigated at a time. (2) The behaviour of such character or characters is determined for each individual separately (i.e. the offspring of each individual is recorded separately). (3) The hereditary constitution of the parents is deduced from the characters exhibited in the offspring and the relative proportions in which these characters appear.

The following examples may help to make the procedure clear.

EXAMPLE I

Parents   Pure Tall Plant   ×   Heterozygous Tall Plant

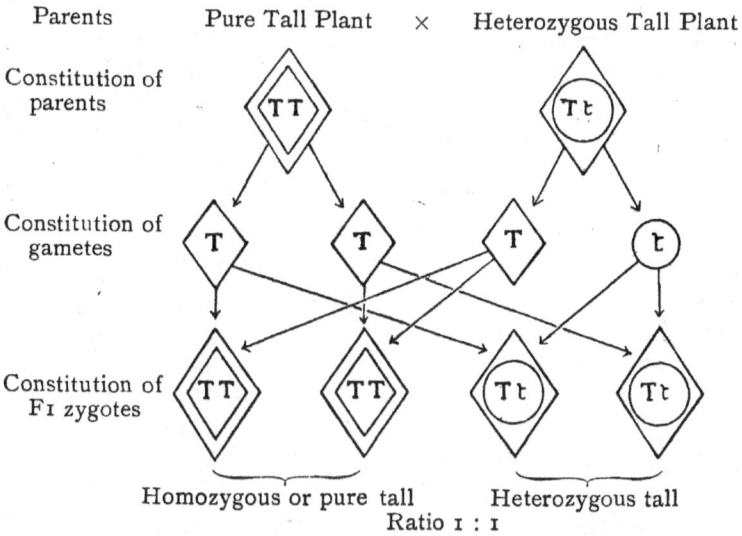

FIG. 28—DIAGRAM TO ILLUSTRATE THE RESULT OF CROSSING
A DOMINANT HOMOZYGOUS PLANT WITH A HETEROZYGOUS PLANT

i.e. the family consists of *tall plants only*, of which half are pure tall and half heterozygous for tallness.

Owing to the fact that the factor for tallness is dominant, all the plants are of tall stature.

EXAMPLE II

Parents   Heterozygous Tall Plant   ×   Pure Dwarf Plant

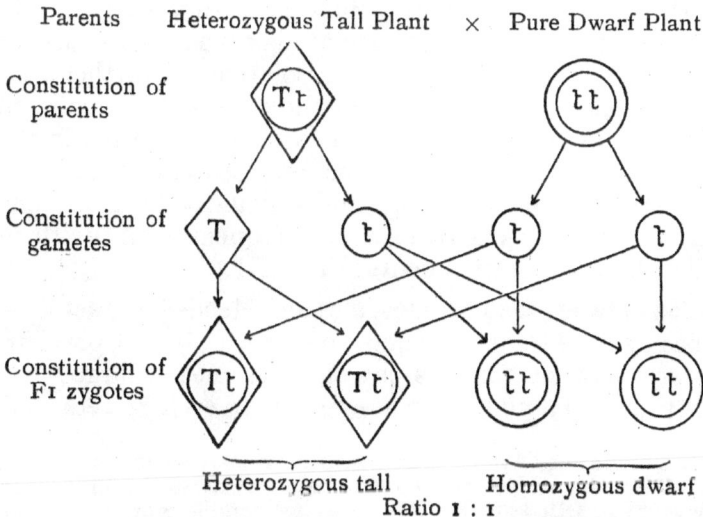

FIG. 29—DIAGRAM TO ILLUSTRATE THE RESULT OF CROSSING
A HETEROZYGOUS PLANT WITH A RECESSIVE HOMOZYGOUS PLANT

i.e. the family consists of *equal numbers of tall plants and dwarf plants*, all the tall plants being heterozygous and all the dwarf plants containing two doses of a factor which behaves as a recessive when crossed.

If the method of dealing with a single pair of factors has been thoroughly grasped there will be no difficulty in applying it to the treatment of two or more pairs of factors simultaneously. It must be remembered, however, that with the inclusion of each additional pair of factors, the calculations become increasingly complex and difficult owing to the larger number of types of gametes formed, and to the greatly increased number of ways in which matings are possible between them. The diagram (fig. 30) illustrates how the calculations for two pairs of factors are made and should enable the student to work out other examples for himself. The plant used is Wheat and the two pairs of characters considered are long and short " awns," and lax and dense arrangement of the flowers in the spike of the inflorescence. Of these, " long awn " is completely dominant over " short awn ": " lax," however, is incompletely dominant over " dense " so that the heterozygote for this factor is intermediate in appearance between homozygous " lax " and homozygous " dense."

It will be noted that if the awned character alone is considered, the proportion of awned to awnless is 12 : 4, i.e. 3 : 1; and if the arrangement in the spike is considered by itself, the proportions of lax, intermediate-lax and dense are as 4 : 8 : 4, i.e. as 1 : 2 : 1.* It may be pointed out further that if there is complete dominance in the case of both pairs of factors, then there will be only four types of offspring, in the proportions of 9 : 3 : 3 : 1, the type represented by lax and intermediate-lax above being indistinguishable from one another. This 9 : 3 : 3 : 1 ratio is one frequently met with in Mendelian breeding experiments (Plate III.).

Relation between the Chromosomes and Mendelian Factors— It has been pointed out elsewhere (pp. 129, 138) that previous to the formation of gametes the chromosomes separate into two groups of equal number, one set going to each gamete. At

---

* Both these ratios have been met with previously when considering the inheritance of a single pair of factors and are those obtained when a heterozygote is selfed in the case of either complete or incomplete dominance.

Parents	Awned, Lax	Awnless, Dense
Constitution	**AALL**	**aall**

Gametes, all of one type	**AL**	**al**

F_1 (all awned, intermediate lax)

AaLl

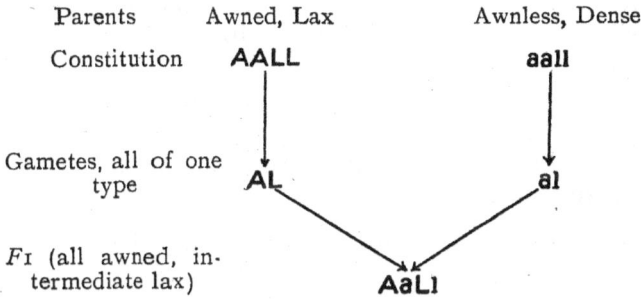

Selfing F_1, we have:—

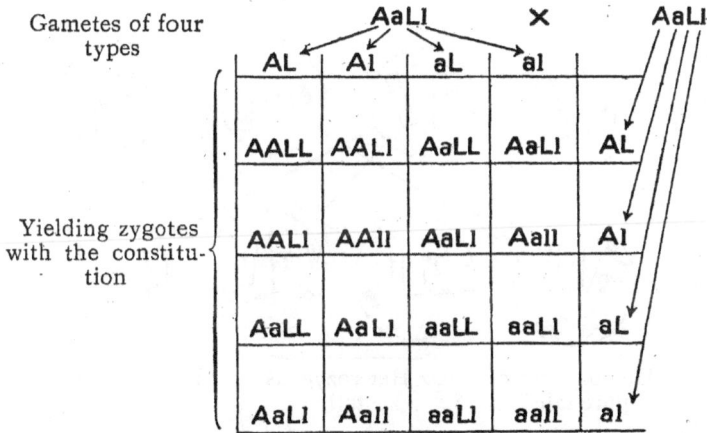

Gametes of four types

AaLl × AaLl

Yielding zygotes with the constitution

	AL	Al	aL	al	
	AALL	AALl	AaLL	AaLl	AL
	AALl	AAll	AaLl	Aall	Al
	AaLL	AaLl	aaLL	aaLl	aL
	AaLl	Aall	aaLl	aall	al

Putting together like terms:—

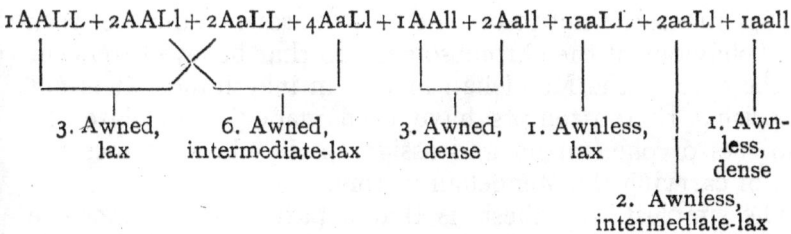

1AALL + 2AALl + 2AaLL + 4AaLl + 1AAll + 2Aall + 1aaLL + 2aaLl + 1aall

3. Awned, lax   6. Awned, intermediate-lax   3. Awned, dense   1. Awnless, lax   1. Awnless, dense

2. Awnless, intermediate-lax

Fig. 30—Diagram to illustrate Mode of Inheritance of Two Pairs of Factors in Wheat

fertilisation each gamete contributes its chromosomes to the zygote, which thus possesses a double set of chromosomes,—a condition which persists throughout the life of the plant until the next " reduction division."

Thus, the vegetative cells of a Flowering Plant possess a double set of chromosomes, and it is believed that every chromosome of the one set pairs with a definite partner of the other set during meiosis (p. 138).

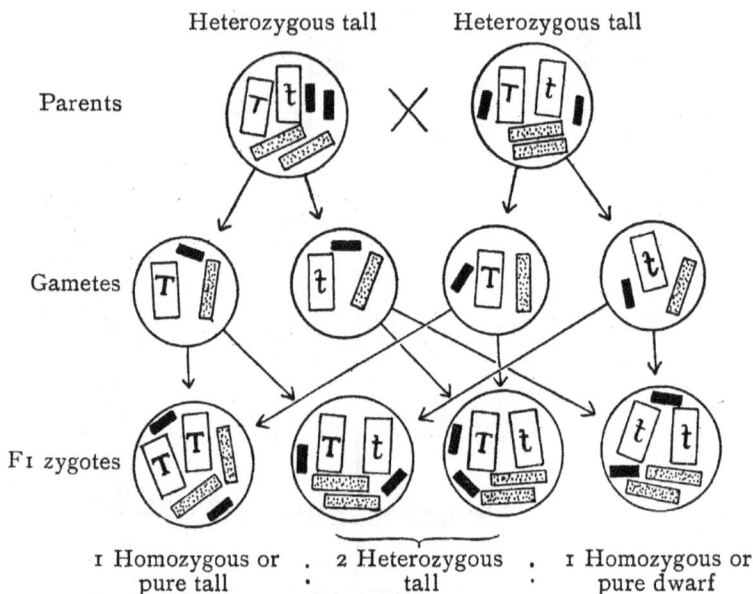

FIG. 31—DIAGRAM ILLUSTRATING THE RELATION CONNECTING THE BEHAVIOUR OF CHROMOSOMES WITH THE FACTS OF MENDELIAN HEREDITY

It is impossible not to be struck by the resemblance between the behaviour of the chromosomes and that believed to occur in the case of the Mendelian factors in inheritance; it is not surprising that attempts have been made to correlate the two sets of phenomena and assign to the chromosomes the rôle of carrying the Mendelian factors.

The simplest hypothesis is that a particular chromosome or its partner carries a particular factor or its allelomorph which can be carried by no other chromosome. How this assumption accords with the results of breeding experiments is illustrated in the diagram (fig. 31).

The diagram represents the crossing of two plants each of which is heterozygous with regard to tallness (T) and shortness (t) of stature. The plants are assumed to possess in each of their nuclei three pairs of chromosomes which are distinguished diagrammatically by representing them as of different shapes. The factors T and t are carried by the chromosomes of the largest pair. The gametes are of two kinds in respect to the factors T and t, and the plants of the F1 generation obtained by crossing are of three types in the ratio 1:2:1. According to this hypothesis, therefore, the known behaviour of the chromosomes is adequate to account for the observed distribution of the characters T and t in breeding experiments.

There is an obvious difficulty, if one factor only is carried by each chromosome, inasmuch as an organism possesses many more factors than chromosomes. The simple hypothesis outlined has been modified, therefore, by the assumption that each chromosome is a compound structure, carrying a large number of factors. It is believed that, during the early stages of meiosis, when corresponding chromosomes come together, interchanges between the pairs of factors carried by each can take place, and so effect a rearrangement. It is possible to imagine a " shuffling " of factors in this way which will give a similar numerical result to that produced if each chromosome were carrying only a single factor.

It is necessary to exercise great caution when seeking to identify two independent series of events, which appear to follow a similar course. This must be borne in mind when we attempt to interpret the behaviour of the unit characters in heredity in the light of the observed behaviour of chromosomes in the course of sexual reproduction. At the same time, the sequence of events in the two cases is so alike that it is difficult not to believe that there is a real coincidence between them. Such an interpretation makes it easier to understand the importance of the nucleus to the cell, and the wonderful precision with which the chromosomes are halved at nuclear division.

The behaviour of a very large number of characters has been investigated by the Mendelian method of studying heredity. In the majority of cases the characters studied are such as distinguish races or varieties of plants from one another. It is seldom possible to investigate the larger differences which distinguish species owing to the difficulties that arise when attempts are made to cross two species, but there is some evidence that the same laws hold at least in some cases.

When any but fairly close relatives are crossed, either no offspring is obtained or it is impossible to carry the experiments further owing to the offspring being sterile. On the other hand, a certain amount of divergence between the parents leads to the production of more vigorous plants in the subsequent generation. We have also in plants the phenomenon of "self-sterility," which means that individual plants are sterile when the flowers are self-pollinated, e.g. Passion Flower, or that they are sterile unless crossed by individuals of a different race or variety, e.g. some of the cultivated varieties of plum and apple.

For the convenience of readers who desire to study the problems of genetics in greater detail, the following list of books may prove useful:

" Heredity." J. A. Thomson. John Murray, 1919.
" Recent Progress in the Study of Variation, Heredity, and Evolution." R. H. Lock. John Murray.
" Mendel's Principles of Heredity." W. Bateson. Cambridge University Press.
" Mendelism." R. C. Punnett. Macmillan and Co.

PART III—THE PLANT IN RELATION TO THE OUTSIDE WORLD

CHAPTER XIII

PLANT RESPONSE

IRRITABILITY; STIMULUS AND REACTION. GENERAL RESPONSES SHOWN BY PLANTS TO CHANGES IN ENVIRONMENT. SPECIAL RESPONSES TO CERTAIN STIMULI BY MEANS OF MOVEMENT; LOCOMOTORY OR BY CURVATURES. TROPISMS. SPONTANEOUS MOVEMENTS. MOVEMENTS INDUCED BY EXTERNAL CONDITIONS. THE PERCEPTION OF STIMULI: "SENSE ORGANS" OF PLANTS. THE TRANSMISSION OF STIMULI FROM A REGION OF PERCEPTION TO A REGION WHERE MOVEMENT TAKES PLACE. MOVEMENTS DUE TO THE STIMULUS OF GRAVITY (GEOTROPISM). THE KLINOSTAT. PRESENTATION TIME AND REACTION TIME. PERCEPTION TIME. MOVEMENTS DUE TO THE STIMULUS OF LIGHT (PHOTOTROPISM). MOVEMENTS DUE TO THE STIMULUS OF OTHER EXTERNAL CONDITIONS. NON-DIRECTIVE (NASTIC) MOVEMENTS. "SLEEP" MOVEMENTS. THE MACHINERY OF PERCEPTION AND RESPONSE. THE "ADAPTIVE" CHARACTER OF MANY MOVEMENTS

The Nature of Plant Response—All living things, whether plants or animals, are characterised by the remarkable property of *responding* or *reacting* to changes in their environment. It is the business of the plant physiologist to study this property in plants, to discover what changes in external conditions provoke a reaction, and to investigate the nature and mechanism of the responses induced.

External conditions may be regarded as affecting plants *directly* through their nutrition and *indirectly* by acting as *stimuli*. In considering nutrition, the student has learnt that plants are influenced at all stages of their existence by the conditions around them. The various processes of nutrition and consequently the rate of growth vary directly, not only with the food supply, but also with changes in external conditions such as temperature or the amount of light. Such fluctuations in growth are usually brought about by changes in the rates at which the various processes of nutrition are carried on, and are of a similar kind to those induced in chemical reactions taking place in the laboratory.

Apart from conditions which act directly in this way, it is convenient to speak of any external condition which produces a direct effect as a *stimulus*, and to refer to any change induced thereby as a *response* or *reaction* to that stimulus. It is not easy to define exactly what is meant by a stimulus, or to trace in detail the connection between some external condition and the response which it evokes. The property inherent in living protoplasm, whereby organisms can show changes of activity in relation to external conditions, has been named *irritability*, but the term is not a very happy one, nor does it help us in any way to understand the real significance of what happens. For our present purpose it suffices to know that this " sensitiveness " is a fundamental characteristic of living matter. It implies that external conditions can provoke changes of a physical or chemical kind in the living protoplasm of the cells, the effect of which is manifested by an alteration of growth or activity in the same or another part of the organism.

It is convenient to consider conditions which act as stimuli under three heads. (*a*) Those which operate by providing an indispensable or *formal* condition for some life process, e.g. a certain minimum temperature is essential in order that a seed may germinate; a certain amount of water must be present for living protoplasm to manifest activity; in these two cases temperature and water act as formal stimuli. (*b*) Those which exercise an effect upon the form of a plant or are *formative* in their action—e.g. the effect of the stimulus of light upon the length of the internodes of a stem. It may be noted that temperature does not usually act as a formative stimulus; thus, while it is possible to tell at a glance whether a plant has been grown in the light or not, it is impossible to deduce from its appearance whether it grows in a hot or a cold region. (*c*) Those which provoke *directive* or other movements; these are considered at length in the present chapter.

Certain external conditions may affect plants in all the ways described, and this is notably so in the case of light.

Light affects green plants *directly* through their nutrition by controlling the rate of carbon assimilation; it acts also as a *formal stimulus*, since chlorophyll is not formed in the dark;* it markedly affects the *form* of plants, and it is responsible for many plant *movements* (p. 197).

Remarkable evidence as to these effects of light can be obtained by growing seedlings in the dark.

Etiolation—Seedlings kept in darkness compare sharply with those grown in light, and the differences may be summarised as follows: In darkness, the green pigments of chlorophyll are not formed, carbon assimilation is prevented, and the plants soon fall behind in growth owing to starvation. The stomata remain closed, transpiration is impeded and the tissues become more succulent owing to retention of water. These symptoms of malnutrition lead in time to general physiological disturbance and ultimately to death.

Light exercises a retarding effect upon growth, and dark-grown seedlings show great elongation of the aerial parts. In Dicotyledons, the stem internodes become greatly elongated: in many Monocotyledons, e.g. Grasses, the leaves increase in length and the stem is not affected. In plants belonging to the former class, the leaves remain small and take up a position nearly parallel to the stem instead of at right angles to it. The stem tip also often remains curved instead of straightening out (p. 199). These peculiarities are described collectively as *etiolation* and a plant showing them is said to be *etiolated*.

In dim light similar reactions occur, differing only in degree from those described. Thus, the gardener places his boxes of young seedlings near the light, so that they may not become " drawn," and plants growing in dense shade are often pale in colour and have long, weak stems.

It is worth noting that the effect of elongation is to bring the shoot into the light and so promote carbon assimilation, e.g. in the case of deeply buried seedlings. The lengthening of the internodes, so characteristic of climbing plants, may be regarded as due to similar causes. Indeed, climbers may be looked upon as plants which have solved the problem of obtaining sufficient light for their requirements in this manner.

* In rare cases chlorophyll is present in the embryo of the resting seed; e.g., Pine, Sycamore.

PLANT MOVEMENTS

The remainder of the present chapter is devoted to a study of the more obvious ways in which plants respond to external conditions by means of *movement*, and a consideration of the nature of plant movements in general.

Most animals are capable of independent movement or *locomotion* and can move towards or away from the objects which surround them. This is the case, also, with some of the simpler plants which live in water and move freely by means of *flagella* or *cilia*. Thus, if we take a little water from a stagnant pond or a water barrel containing the minute green cells of the alga *Chlamydomonas* and place it in a saucer on a table near a window, we find that all the cells move to one side of the saucer, leaving the rest of the water clear. If we carry the experiment further, we learn that the position taken up by the algal cells depends upon the stimulus of light. Similarly, if small amounts of various chemical substances are added to the liquid in which free-moving plants (e.g. Bacteria) are growing, the cells may be observed to move towards or away from the regions of greatest concentration.

In the case of movements depending upon the effect of light, the position taken up by free-moving green plants is usually that most favourable to carbon assimilation, and the power of responding in this way to the action of light brings the cells into a better position for carrying on their vital activities and is, therefore, in the nature of an *adaptation*. This is by no means always the case, however, and such cells, if influenced by the presence of chemical substances, may move towards poisons, i.e. towards their own destruction, as readily as towards substances of a beneficial kind.

In the Higher Plants, the plant is fixed by its roots in the soil or elsewhere, and movement as a whole, towards or away from the objects around it, or in relation with external conditions generally, is impossible. Movement of living parts is therefore limited to *elongation* and *contraction* of tissues due to changes in the degree of turgor of the cells or to growth, and to *curvatures* of different organs resulting from inequalities of turgor or of the rate of growth on different sides of the organ.

" Spontaneous " Movements—Not all plant movements can be related directly to changes in the environment. Certain movements in plant cells or organs are apparently spontaneous, i.e. they do not appear to be induced by an external stimulus. It is customary to describe such movements as *spontaneous* or *autonomous* and thus to distinguish them from those related directly to an external cause or stimulus,—*induced* movements as the latter are called. It is not impossible, however, that the distinction is an artificial one, serving to conceal our ignorance of the real nature of the immediate causes of movements belonging to the former class.

Examples of such " spontaneous " movements are the remarkable streaming movements of the protoplasm in many plant cells, and also movements of the nucleus and other cell-structures from one part of the cell to another (p. 17). Certain of the lower plants, too, show curious, spontaneous, creeping movements, the actual mechanism of which is not very clearly understood.

In the Higher Plants, growth in length of the stems and roots results from the activity of the growing points. The forward movement of roots, shoots and other organs as a consequence of growth does not take place in a straight line. This can be verified in the case of stems, when a growing shoot is observed from above under conditions such that no external stimulus tends to evoke a curvature. Under such conditions the stem tip will be found to advance in an irregular spiral. This movement during forward growth is called *nutation* or *circumnutation*. It is due to growth taking place more rapidly on one side of the tip and to a continuous change in the position of this region of most active growth, and apparently depends upon some property inherent in the plant rather than upon external causes.

A remarkable case of what is apparently spontaneous movement is that shown by the leaves of the Telegraph Plant (*Desmodium gyrans*), a native of India. This plant has compound leaves each with three leaflets. As in many other plants belonging to the Pea and Bean family (*Leguminosæ*), the leaves show " sleep movements " of a kind similar to those exhibited by clover and other common plants. In addition to this the leaflets, more especially the two lateral leaflets, have regular movements which apparently take place spontaneously and which, so far as is known, serve no special purpose in the general economy of the plant. In the course of the movement the tip of each leaflet describes an ellipse, taking about $1\frac{1}{2}$ minutes to return to its original position. Movements similar to this but more difficult to detect are shown also by the leaves of other plants.

Movements Induced by External Conditions—Movements in water of free-swimming plant cells, depending upon the stimulus of light, have already been mentioned. In the Higher Plants, movements of this kind take the form of *curvatures*, by means of which the various organs of the plant, or distinct parts of these organs, take up different positions which determine the general *habit* of the plant. These positions are not assumed at random nor are they the result of chance. They are a consequence of responses shown during growth to external conditions which act as stimuli.

It must be noted that certain external conditions act as *stimuli* only if the plant is placed so that the organs are in positions different to those normal to them, or the *effect of the stimulus only becomes evident* when the organs are so placed. Moreover, different organs respond differently and in varying degree, and the positions finally assumed may be regarded as the *resultant* of external and internal stimuli acting simultaneously. The nature of the stimulus and the response which it evokes can only be determined by experiment.

In the case of animals we are accustomed to the idea of special sense-organs, by means of which the animal " perceives " external stimuli such as those induced by light, sound or gravity. In the lower animals such sense-organs are often rudimentary and are represented by parts of the body specially sensitive to certain stimuli or impressions. In many plants it is also found that the power of " perceiving " external stimuli is limited to certain regions which may fairly be compared with the simple sense organs found in the lower animals.

In plants, as in animals, the power of responding to an external stimulus by movement is often located in a different part of the body from that which acts as a sense organ. Sometimes the two regions are close together but often this is not so, in which case the " message " received in one part may have to be transmitted over a considerable distance.

In the higher animals there is a wonderful telegraphic

system,—the nervous system,—by means of which impressions or messages received in one part are rapidly transmitted
to the central nervous system—the spinal cord and brain,—
from which another message is sent to the muscles that move
the limbs.

Since there is nothing in plants which can be compared
structurally with the nerve of an animal, it is presumed
that the stimulus travels from one part to another by way
of the protoplasm, the continuity of which is ensured by the
fine threads or *plasmodesmen* which penetrate the walls of
adjoining cells.*

The most marked reactions of plants to external stimuli
depend upon *gravity* and *light*. In the case of Flowering
Plants, for example, the direction of growth of the roots and
shoots and the positions taken up by these organs when
mature, may be regarded as due mainly to the combined
effects of light and gravity acting together or in opposite
directions. Such movements, resulting in the placing of
organs in positions which show definite relations to the direction in which the stimulus is acting, are called *tropic* movements or *tropisms*.

Movements and curvatures are also induced by stimuli
due to *moisture, injury, contact*, and other external conditions.
The nature of some of these reactions will now be briefly considered. The statements about to be made should be verified
by carrying out the experiments at the end of this chapter
and extended by performing others of a similar character.

Movements Due to the Stimulus of Gravity (Geotropism)—
Gravity is the force attracting all bodies to the centre of
the earth. It affects plants not only in this direct way,
but also indirectly as a stimulus, the reactions to which
influence the direction of growth of the various organs. This
is true not only for the Higher Plants, but also for Ferns,
Mosses, Fungi, and Algæ. A few plants, e.g. the Mistletoe,
are apparently indifferent to the stimulus of gravity, in so far
as it affects the direction of growth. The nature of the
response varies with the kind of plant and the part of the
plant stimulated.

In the case of fixed plants, the organs stimulated by

* A special transmitting tissue has been identified in the case of
Mimosa pudica.

gravity usually show curvatures, tending to bring them into definite positions with reference to the *direction* in which the force is acting—i.e. to a line passing through the centre of the earth. The movement is, therefore, described as a *tropism*, and this particular aspect of plant response is known as *geotropism* (Greek: γῆ, the earth, τροπή, a turning).

Different parts of the plant react in different ways, as is illustrated by the following examples, in which the action of other stimuli is assumed to be eliminated.

If a young mushroom be laid on its side, the stalk will curve until the " umbrella " or *pileus* is in the original position. The root of a young seedling emerging from the seed grows vertically downwards in whatever position the seed is planted; the young shoot grows vertically upwards; the lateral roots and lateral shoots place themselves at definite angles with the vertical. These facts are described by saying that the seedling root is *positively geotropic*, the seedling shoot and the stalk of the mushroom *negatively geotropic*, while the lateral organs show *lateral geotropism* or are *diageotropic*.

It is significant that these reactions are of value to the plant in each case, inasmuch as they result in the several organs taking up positions favourable to the work which they perform, and in this way tend to overcome the disadvantages incidental to a fixed position.

Thus, the mushroom is placed in a position such that spores can fall clear of the gills and have a good chance of being wafted to some distance by the wind before they reach the ground. The seedling root is carried down into the soil, whence it obtains supplies of water and salts, and the lateral roots are so distributed that a considerable area of soil is explored. The seedling shoot is carried up into the light and air,—conditions essential for the proper performance of its functions.

The Klinostat—A plant organ, e.g. a young root or shoot, not subject to any stimulus which provokes curvature, grows equally, although not necessarily simultaneously, on all sides (p. 189). If one side is stimulated or if the position is altered in such a way that a stimulus, previously acting equally on all sides, acts more strongly on one side, growth will take place unequally and a curvature result.

In the case of geotropic curvatures, experimental conditions are rendered difficult because it is impossible to remove the plant from the operation of the stimulus. It is, however, possible to arrange conditions such that the stimulus of gravity acts successively on every side of the organ concerned, with the result that the tendency to curve in one direction is immediately counteracted by that to curve in another and so on. The stimulus of gravity is not thereby removed, but is distributed equally on all sides of the organ, which accordingly continues to grow straight forward.

To produce this experimental condition a simple apparatus called a *klinostat* is required. It consists of a piece of clockwork, to which is attached a rod or spindle which revolves when the clockwork is set in motion. A pad of moistened cotton-wool or other suitable material to which seeds can be attached is fitted to the spindle and the latter placed in a horizontal position in a dark chamber. If the axis is at rest the roots of the seedlings, as they emerge, show positive geotropism and curve downwards, the shoots show negative geotropism and curve upwards. If the experiment is repeated with the difference that the horizontal axis of the klinostat is kept revolving, the roots and shoots of the seedlings, in whatever direction they emerge, continue to grow straight forward. The first experiment acts as a *control*, i.e. reproduces the experimental conditions except with regard to the particular condition under investigation,—in this case the effect of gravity on a seedling axis.

Special attention should be paid to the general conditions under which such experiments are carried out; e.g. they should be set up in the dark because many parts of plants which react to gravity are also sensitive to the stimulus of light and the effects of the latter might obscure those of the former. It is important, also, to insure that the temperature is favourable and that sufficient moisture is supplied.

Early students of geotropism discovered that the effects of gravity could be neutralised by growing plants on a rapidly revolving wheel placed in either a vertical or a horizontal position; i.e. they found that the effect of gravity could be overpowered by that of *centrifugal force* when the two forces were acting in opposition to one another. When the two forces are not acting in direct opposition, the position assumed by any sensitive, curving organ is a "resultant" of the combined effects.

Presentation Time and Reaction Time—In order that a root should show a geotropic curvature it must be exposed to the stimulus of gravity for a measurable period, the minimum length of which can be determined by experiment. A measurable time must also elapse after "perception" of the stimulus before the curvature begins. These two periods are known respectively as *presentation time* and *reaction time*.

It is interesting to learn that the former can be made up of a series of exposures, the sum of which is equal to the time required. Thus, if the root of a seedling bean requires twenty minutes' exposure in a horizontal position in order to show a geotropic curvature, it suffices to place the root in a horizontal position during four periods of five minutes each, in the intervals between which the root is unstimulated, e.g. it is placed in a vertical position or on a horizontally revolving klinostat.

Perception Time—Each of the periods during which the root is exposed must be long enough for the stimulus to be effective, i.e. for the root to "perceive" it. The shortest period which suffices for this is called the *perception time*. Furthermore, the intervals between the periods of exposure to the stimulus must not be too prolonged or the effect of one period of stimulation will disappear before the next is received. The following simile, due to a distinguished plant physiologist, may make the matter clearer.

Compare the seedling with a gun, fired by means of a trigger. The root curves in response to a stimulus; the gun is let off by application of a weight to the trigger. Let us imagine, further, that the weight which releases the trigger is composed of a quantity of water, collected in a vessel of some sort hung on the trigger. Sufficient water may be supplied all at once, or it may be added drop by drop. If the latter method is chosen, and the interval between successive drops is sufficient for each to evaporate before its successor is added, then the gun will never be fired. If, on the other hand, the drops follow one another at such short intervals that the loss from evaporation is negligible, then the reaction will follow as soon as sufficient weight of water has accumulated.

The Localisation of Perception—It has already been suggested that plants possess organs comparable with the sense organs of animals. This is well illustrated in the behaviour of seedling roots.

If the root of a bean seedling is marked with India ink at regular intervals and placed horizontally, it can readily be observed that the *region of curvature* coincides with the *region of growth*. If the extreme tip is removed from each root at the beginning of the experiment, leaving the growing region intact, no geotropic curvature takes place, although the root continues to grow. In a bean root, therefore, " perception " of gravity is *localised* in the extreme tip, which serves as a " sense organ " for this particular stimulus. It follows that the stimulus perceived must travel or must be transmitted from the tip to the growing region behind it, in order that a curvature may take place.

Experiments which involve the cutting or wounding of living tissues must be performed with the greatest care,—must be treated, in short, as surgical operations. The experiment just described is open to criticism on the grounds that roots are sensitive to the stimulus of wounding as well as to that of gravity and that the results of the experiment are thereby obscured. This criticism is a just one, but the injury due to wounding can be reduced to a minimum, and the evidence that " perception " of the stimulus of gravity is localised at the tip of the root is amply confirmed by more difficult experiments which do not involve removal of the tip. Further inquiry into these matters, as also into the nature of the mechanism by which plants can " perceive " the stimulus of gravity, may be postponed until the student has reached a more advanced stage of the subject.

Another instructive experiment to illustrate the separation of the " perceptive " from the " reacting " regions in plants can be performed with the seedlings of certain Grasses in which the first leaf or " cotyledon " is sensitive to gravity while the curvature occurs some way from the tip.

If such a seedling, growing in a pot, is placed in a horizontal position, the stimulus " perceived " by the " cotyledon " is transmitted to the reacting region, with the result that the lower side in the latter grows more rapidly than the upper side and a curvature results which brings the " cotyledon " into the unstimulated (i.e. vertical) position.

Suppose, now, that instead of a seedling growing in a pot we take one which has been germinated in moist air and fix it in a horizontal position by thrusting the " cotyledon " into a narrow glass tube, leaving the rest of the seedling free to move. The " cotyledon," as before, perceives the stimulus of gravity and the lower side of the reacting region grows more rapidly producing a curvature. In this case, however, curving does not bring the " cotyledon " into the unstimulated position, so that it continues, and may result in the stem becoming twisted into a series of coils (fig. 32). Thus, although curvature under the influence of a stimulus ordinarily

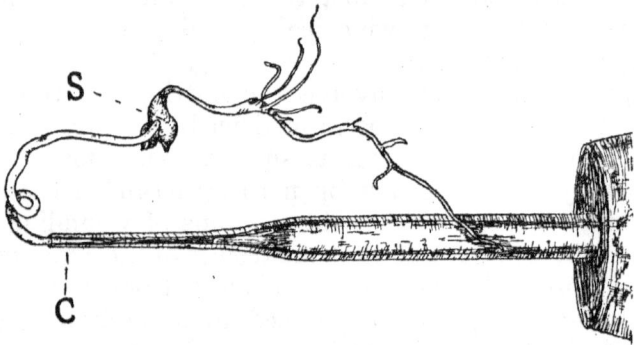

FIG. 32.—GEOTROPIC RESPONSE IN A GRASS SEEDLING

The cotyledon, *c*, of a Grass (*Sorghum sp.*) is held horizontally by inserting it in a glass tube. The stimulus of gravity is transmitted from the " region of perception " at the tip of the cotyledon to the " region of curvature " behind it, which continues to curve as shown in the drawing; *s*, seed

tends to bring the part displaced back into its normal position, yet the reaction described is not really " purposeful," but is merely a blind response to a particular stimulus.

The power of responding to the stimulus of gravity is not confined to the growing regions of stems, roots or other organs; more rarely it is shown by tissues of mature parts such as the nodes of the stems of Grasses and other plants (p. 204).

Finally, reaction to the stimulus of gravity may be shown in plants by the formation of structures or tissues in special positions. Such reactions, although of great interest, cannot be considered in detail (p. 204).

Movements Due to the Stimulus of Light—The very existence of green plants is bound up with " a place in the sun " and it is not, therefore, surprising to find that they show marked responses to the effect of light, many of which take the form of movements or curvatures bringing the plant or certain of its organs into positions favourable for photosynthesis.

The stimulus of light acts upon green plants in two ways. In the first place, it determines the direction of growth of different organs; in the second place, it exercises a general effect in the absence of which certain movements or curvatures do not occur or take place very imperfectly. A majority of curvatures induced by light are of the nature of tropisms and are known as *phototropic* or *heliotropic* movements.

The general effect of light is well shown by the curvatures induced if a pot plant of e.g. Geranium or *Tropæolum* is placed in a position where the light falls upon it from one side only. The stem curves in the direction from which the light is coming,—it is *positively phototropic*. The leaves place themselves so that their flat surfaces are at right angles to the rays of light,—they are *diaphototropic*. If a seedling is used for the experiment, it can also be determined that the young root curves away from the light,—it is *negatively phototropic*.

Whilst the positions taken up by the roots and shoots of an ordinary plant are in the main determined by stimuli due to light and gravity, the effect of gravity is more marked upon the roots and underground parts and that of light is more marked in the case of the aerial shoot system. Reactions due to light and gravity respectively must be carefully distinguished by means of experiments. In the case of the former, the effect of the stimulus can be avoided by placing the experimental plant in the dark; in the case of the latter, the effect can be equally distributed and rendered ineffective by performing the experiment upon a moving klinostat. (The klinostat can also be used for distributing the effects of one-sided illumination.)

Many underground stems show light reactions similar to those of roots, and many interesting differences in behaviour may be noted with regard to the positions taken up by the leaves of different plants.

The Perception of Light Stimulus—It can be shown experimentally that " perception " of light, like that of gravity, is sometimes localised to regions situated at some distance from those which curve as a result of the stimulus.   In the leaves of some plants, for example, it has been shown that " perception " of the light stimulus is limited to the leaf-blade, whereas the stalk or petiole is the part which responds by means of a curvature.   One of the best known cases is that of the first leaf of certain grasses, in which " perception " is localised to the extreme tip.

It is usual for the leaves of Monocotyledons (to which group the Grasses belong) to exhibit *positive phototropism* —i.e. they curve so as to place themselves parallel to the direction in which light is acting.   If, therefore, a pot containing seedlings of e.g. Millet (*Sorghum sp.*) is placed in a dark box into which light is admitted through a window in one side, all the seedling leaves bend towards the window, the region of curvature being situated some distance behind the tip.   If the experiment is repeated, having first carefully " blinded " the seedlings by covering the tips with little caps of tinfoil, the leaves show no reaction because they are unable to " perceive " the stimulus.

Movements Due to the Effect of Other External Stimuli— Movements and curvatures are also shown by plants in response to other external stimuli, notably to water (*hydrotropism*), chemical substances (*chemotropism*), or contact (*thigmotropism*).   Examples of such responses are given at the end of this chapter.

Many of the reactions induced by contact are well known to everyone and were among the earliest cases of plant movement investigated.   Among such are the sudden movements shown by the stamens of some flowers when touched, (Barberry, Cornflower), the curious movements of the " trap-leaves " of many insectivorous plants (Sundew, Venus' Fly-trap) and the sensitiveness to contact shown by many tendrils (Passion Flower, Bryony).

Non-directive (Nastic) Movements—Reactions to stimuli are sometimes shown otherwise than by tropisms.   An example of such is the one-sided growth which causes the straightening out of the young shoot of a Bean or other seedling after it

PLATE IV

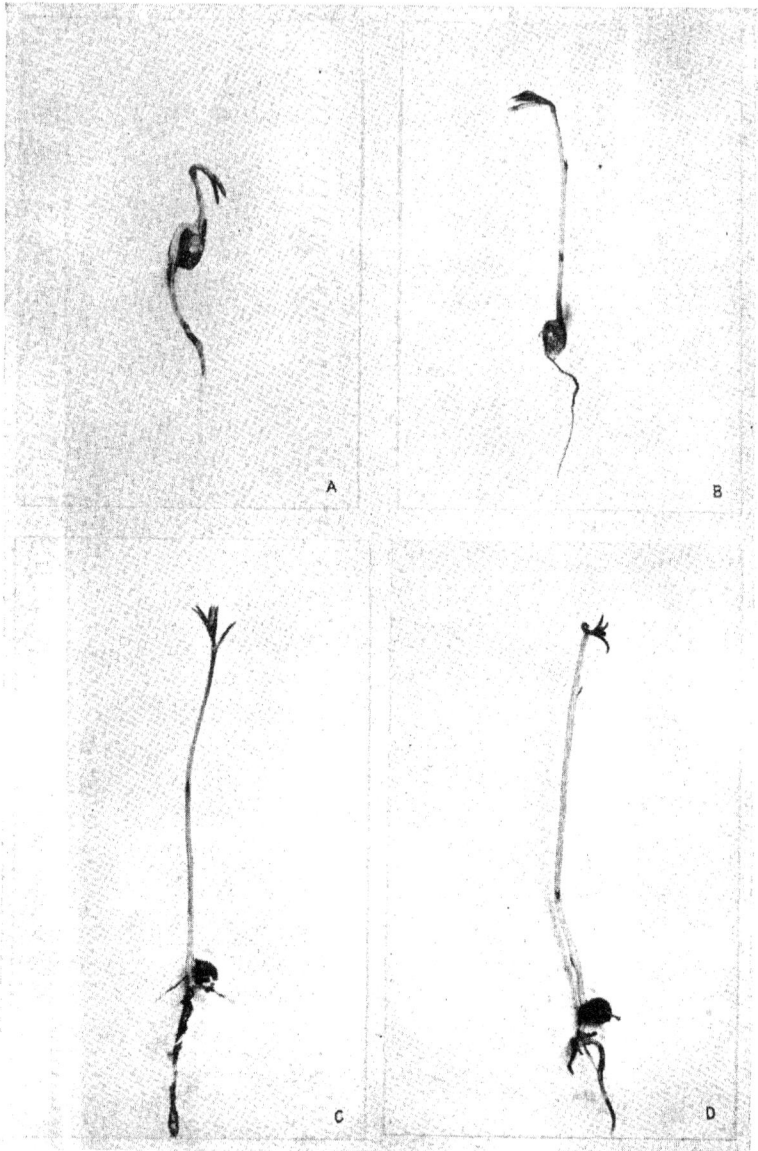

CURVATURES DUE TO THE STIMULUS OF LIGHT

A, seedling of Vetch (*Vicia sativa*) germinated in the light. B, C, same seedling at later stages of growth, showing straightening of the shoot in the light. D, a similar seedling germinated and grown in the dark. The unequal growth causing the curvature of the shoot apex shown in A has persisted and resulted in the formation of a coil by the stem apex.

reaches the light. The hook-like appearance of the plumule is due to more rapid growth on the side which is uppermost as it pushes through the soil. On reaching the light, this inequality of growth is reversed; elongation takes place more rapidly on the under side and the apex of the shoot straightens out into a vertical position. This change in the distribution of growth depends on the stimulus of light, since the seedlings of some plants, if germinated and grown in the dark, continue to show more rapid growth of the upper side, so that the young stem grows in a series of coils; those of other plants straighten slowly and imperfectly if kept in the dark. The reaction evidently depends on the stimulus of light, but the degree of dependence on the stimulus varies with the plant (Plate IV.). Of similar nature are the opening and closing movements of many flowers, e.g. Tulip, *Crocus*, " Poor Man's Weather-glass " (*Anagallis*), and many others. These movements are due either to changes in the light intensity or to changes of temperature, more frequently to the effect of the latter.

A non-directive movement or curvature of this nature is sometimes described as a *nastic* movement to distinguish it from a directive movement or *tropism*. Such nastic movements are induced by other external stimuli as well as by light.

Sleep Movements—Everyone is familiar with the curious automatic movements shown by the leaves of certain plants in the mornings and evenings, and is aware that in Clover, Wood Sorrel, *Acacia*, *Mimosa* and other plants, the leaflets and leaves take up definite " night " and " day " positions. These so-called " sleep movements " depend upon light and darkness, as may be readily learned by covering a plant with a dark box during the day. They may be classed as nastic movements, and differ from the curvatures hitherto described in that they affect mature organs and are not due to inequality of growth. They depend upon *temporary* changes in the distribution of turgor of the cells of little swellings or *pulvini* at the bases of the leaves and leaflets. The " night position " can be induced artificially by placing the plant in darkness or by covering it with a dark box.

An interesting feature of these movements is that they may persist for some time in the absence of the appropriate stimulus. Thus, if a plant of *Acacia* or Clover is kept in the dark, it continues to attempt to " go to sleep " at night and to " awaken " in the morning for some days. It is as if the protoplasm of the cells which undergo change continues to " remember " the stimulus after its effect has ceased to operate (Plate V.).

The " Sensitive Plant "—The " Sensitive Plant " (*Mimosa pudica*) is a common annual weed in certain parts of the tropics, and can easily be raised from seed in a greenhouse in this country. It is a small plant with doubly compound leaves, and the habit of a miniature *Acacia*. Swellings or pulvini are present at the base of the main leaf stalk, at the bases of the secondary leaf-stalks and at the bases of the leaflets, and are associated with independent movements of these parts. Movements are immediately induced as a result of (1) shock or contact, (2) change of temperature or illumination, (3) chemical stimuli. As a result of shaking or other mechanical stimulation, the application of heat, electrical shock, or the action of irritant vapours like ammonia, the leaflets and secondary leaf-stalks rapidly fold together and the main leaf-stalks fall through a large angle. The leaves and leaflets also show well marked " day " and " night " positions; the " night " position is similar to that assumed as a result of shock, etc. except as regards the main leaf-stalk, which moves upwards instead of downwards. Not only does the plant exhibit sensitiveness of an extreme kind, but it also shows a very remarkable power of transmitting stimuli over long distances. Thus, if a lighted match is held under the tip of a leaflet, the effect may be observed to travel progressively to the other parts of the leaf, and, if the stimulus is sufficiently great, to extend to adjoining leaves.

The case of *Mimosa* must be regarded as exceptional as regards the degree of sensitiveness exhibited. This high degree of sensitiveness does not appear to serve any very obvious useful purpose to the plant.

The Machinery of Perception and Response—It is not possible to deal here with the interesting researches and experiments which have been carried out to investigate the actual machinery whereby stimuli of various kinds are " perceived " by living plant cells, and changes of position induced in the different organs. With regard to the former, it is usual to speak of the " sensitiveness " of the plant, or of some limited region of it,

PLATE V

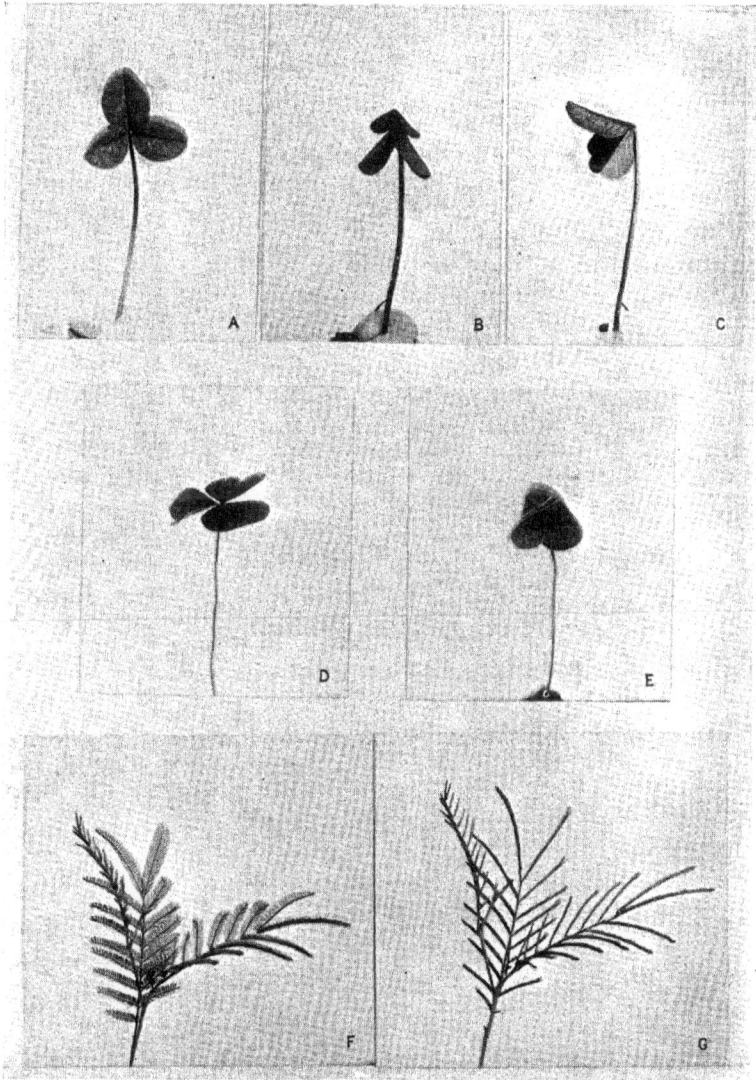

SLEEP MOVEMENTS

A, Leaf of Clover (*Trifolium repens*). B, C, Same leaf in ' sleep' position. D, Leaf of Wood-sorrel (*Oxalis acetosella*). E, Same leaf in ' sleep' position. F, Twig of *Acacia* (*A. aealbata*). G, Same twig with leaves in ' sleep' positions.

to a particular external condition which acts upon it. It is possible to make a rough comparison with the printing paper, sensitive to light, which is used by the photographer. In the latter case, " perception " of light by the paper causes a chemical reaction, the effect of which is made evident as a change of colour. In the case of cells " sensitive " to a stimulus, some change,—possibly a chemical one,—takes place, directly or indirectly, as a result of the stimulus, and this operates by temporarily altering the properties of the cell protoplasm. Such changes can be communicated from one cell to another and the effect of a stimulus thus transmitted for some distance. When response to a stimulus takes the form of a curvature, it is usually shown by young and actively growing parts, and results from an unequal distribution of growth. Movements induced in mature organs are often due to temporary local swelling or shrinkage, depending upon changes in the turgor of the cells. Nothing is known with precision of the actual mechanism whereby the cell-changes induced by " perception " are related to those causing movement. Something is known, however, as to the nature of the changes induced in cells responsible for the " perception " of stimuli due to light, gravity and contact, for an account of which the student must refer to a more advanced textbook.

Attention may be drawn once more to the interesting fact that different parts of the plant give different responses to the same external stimulus, and it can be shown that this depends to some extent upon the great principle of correlation which links up and co-ordinates all the organs of the plant into one organic whole. Thus, the main root of a Bean seedling is positively geotropic, the lateral roots are laterally geotropic; if the main root is cut off, there may follow a change of reaction on the part of one of the lateral roots which will exhibit positive geotropism and so replace the main root in position. In yet another way does the irritability of protoplasm differ from the sensitiveness shown by photographic paper to light. In the case of tropic curvatures depending upon the effect of e.g. gravity or light, the stimulus only becomes evident if the sensitive organ of the plant is placed in an abnormal position with respect to that particular condition.

It must be clearly understood that use of the terms " perceive," " perception," etc. in relation to plants is not intended to suggest a direct comparison with the more complex responses to stimuli exhibited by the higher animals.

Movements Dependent upon the Mechanical Structure of Tissues—The distribution of thick-walled and thin-walled tissues in plant organs is often such as to cause movements when the tissues lose water. In the leaves of the Marram Grass (*Ammophila arenaria*), for example, there are patches of thin-walled cells towards the upper surface of the leaf while the majority of the thick-walled tissue is towards the under surface. The result of the arrangement is that when the amount of water in the leaf becomes reduced from any cause, the thin-walled cells shrink in size while the thick-walled cells retain their size and shape. The unequal contraction on the two sides of the leaf causes the latter to curl along its whole length and so incidentally checks further loss of water from the stomata on the upper surface. Renewal of the water-supply results in the thin-walled cells becoming once more turgid, as a result of which the leaf uncurls.

Movements of an *explosive* character sometimes occur as a result of the development of regions of comparative weakness combined with a high internal pressure in the cells. The seeds of the " Squirting Cucumber " and the spore-containing body or sporangium of the common little fungus *Pilobolus* are distributed in this manner.

In a different class from movements and curvatures of living plants or plant organs are *mechanical* movements brought about by pressure or tensions set up in dead tissues. Most of these are hygroscopic movements brought about by unequal swelling and shrinkage, due to local differences in the capacity for absorbing and retaining moisture. The cells of structures showing such movements often possess conspicuously thick cell-walls, and the nature of the movement by which the tension is relieved is directly related to the chemical nature of these walls and to the distribution of the thickened areas.

Mechanical movements of this kind are frequent in fruits, where they are often related to the dispersal of seeds. In many capsules, for example, tension is set up in the walls of the ripe fruit owing to the unequal loss of water during ripening; this is relieved by a sudden splitting of the wall of the fruit, the seeds are jerked out and may be flung to a considerable distance. The arrangement in fruits is usually such that the tendency to split comes about when the air is dry, when the process can often be initiated by a touch. In some desert plants a reverse condition exists and hygroscopic movements of the fruit or of its appendages allow the seeds to escape only when the air is moist, i.e. when the conditions are most favourable to germination. Common examples of explosive fruits are the common Broom (*Cytisus Saro-thamnus*), and the Violet. Such hygroscopic movements are sometimes reversible, i.e. the parts contract or expand according to the amount of humidity in the air, which, in turn, causes unequal contraction or expansion of different parts of the hygroscopic tissues. A good example is the " self-burying " fruit of the Stork's Bill (*Erodium cicutarium*). In this plant, the ripe fruit splits into five pieces, each of which contains a single seed and possesses a long strip of tissue or *awn* which shows hygroscopic movements when alternately dried and wetted. The repeated movements of the awn as it absorbs water or loses it, according as the air is moist or dry, may tend to bury the part of the fruit containing the seed, and thus to place it in the best position for successful germina-tion. The presence of hairs on the wall of the fruit may assist in the process.

It is impossible to offer any entirely satisfactory explana-tion of the " adaptive " character of so many of the move-ments described in this chapter. The " use and disuse " theory by which Lamarck sought to explain animal structures is not so easily applied to plants, and it is not always possible to give a consistent explanation based on Natural Selection. To quote Sir Francis Darwin:—" the plants alive to-day are the successful ones who have inherited from successful ancestors the power of curving in certain ways, when, by accidental deviations from their normal attitude, some change is produced in their protoplasm."

Practical Work

Apparatus—If a klinostat is not available, one can be improvised by using the works of an ordinary small American clock. A working watch-maker can attach a light metal rod to the spindle which ordinarily carries the minute hand. A piece of moist clean peat, or a pad of moistened cotton-wool, fixed to the end farthest from the clock serves for the germination of seeds; better than either peat or cotton-wool is a piece of " loofah," cut to convenient size, the larger cavities filled with moistened material. Simple experiments on geotropism can also be carried out by the aid of a water-wheel, kept revolving by means of a stream of water from a tap. For more advanced work it is desirable to have a klinostat designed for the purpose. This should be adjustable to revolve at several different rates. It is also advisable to procure a box painted black within to serve as a dark chamber in which germination and other experiments can be kept under moist conditions. One side of the box should open with a shutter; in another side a hole should be cut to admit the axis of the klinostat, the clock-work part of which should be kept outside. The same or a similar box will be useful for experiments upon the effect of light: for this purpose the shutter can be furnished with a window into which a pane of white or coloured glass can be inserted at will.

Geotropism—1. Germinate seeds of e.g. Broad Bean, Pea, Sunflower, or Cress on the klinostat. Arrange the experiment so that the germinating seeds are placed in moist air in a dark chamber and note the behaviour of the seedlings, (a) when the seeds are placed on a horizontal axis at rest, (b) when the axis is kept revolving by means of a klinostat. Repeat the experiment, using pot plants of various kinds.

2. Germinate seeds of Vegetable Marrow in a pot and observe the development of the little " peg," the presence of which facilitates the escape of the seed-leaves from the seed when it is placed flat. Compare the behaviour of seeds planted in an " edgeways " position. Germinate similar seeds on a revolving klinostat and note the effect of *distribution* of the stimulus of gravity on the formation of the peg. The position in which it appears is evidently determined by geotropic response in the first instance.

3. In order to demonstrate that the stalk of a mushroom or toadstool is sensitive to gravity, lay it on its side in a dark box. Repeat the experiment, placing another mushroom or toadstool on the axis of a revolving klinostat, and thus confirm the geotropic character of the reaction.

4. Determine the reaction to gravity of the haulms of a grass. Collect pieces of stem with several nodes. Place them horizontally in a dark box. (A small biscuit-tin serves well for the purpose, and the stems can be fixed horizontally by sticking the ends into a bank of damp sand at one end of the box.) Compare the behaviour of these with that of similar pieces attached for the same length of time to a revolving klinostat.

5. Determine the kind of geotropic reaction shown by (a) the axis of a dandelion head, (b) the peduncle of *Narcissus* or Snowdrop. In each case place one flower or flower-head in a horizontal position in the dark and note any curvatures which take place. Repeat the experiment on a revolving klinostat in the dark and deduce the effect of gravity in causing the curvature.

An instructive experiment can be performed by fixing a grass haulm to a piece of board or sheet cork so that the former is kept in a horizontal position for some considerable time forcibly. Under these conditions, the increased growth on the lower side of the leaf-sheath, by which erection of the prostrate shoot is normally brought about, becomes evident as a swelling on the lower side of the node. The board to which the grass is fixed should be grooved beneath the node to allow the outgrowth free development.

Phototropism—6. Place a pot plant in a dark box, lighted from the side, and note the positions taken up by the stems and leaves. The result can be verified as due to the effects of light by repeating the experiment with the plant rotated on a klinostat placed so that its axis is vertical and the plant is therefore illuminated successively on each side.

7. Repeat experiment 6, using a seedling growing in water or culture solution in a glass vessel, and note the reaction to light shown by the roots.

8. Sow seeds of Canary-grass (*Phalaris canariensis*), Italian Millet (*Setaria italicum*) or Millet (*Sorghum vulgare*) in a pot and note the reaction shown by the first leaf (cotyledon) to one-sided illumination. Note especially the *position* of the curvature.

Localisation of Perception—9. Germinate a number of Broad Bean seeds on moist blotting-paper. Place some of the seedlings horizontally on a stationary axis, others on a horizontal axis kept revolving by means of a klinostat. In each case remove the *extreme* tip with a *sharp*, *clean* knife or scalpel which has been passed through a flame. Compare the behaviour of these with that of the uninjured roots.

10. Prepare a pot of seedlings of Millet as described for experiment 8. Before subjecting them to one-sided illumination, " blind " certain of the seedlings by placing little caps of tinfoil over the tips of the leaves. The caps can be made from ordinary silver paper of good quality by rolling the paper carefully round a piece of glass or metal of similar size to the tip of the leaf. They must be absolutely light-proof and should not be so heavy as to weigh down the leaf.

Presentation Time—11. Place a Bean seedling in a horizontal position for a short time, e.g. fifteen minutes, and then transfer to a klinostat, or place in a vertical position so that it will no longer be stimulated. If the exposure has been sufficiently long, the appropriate curvatures will take place. If no curvatures result, repeat the experiment, exposing the seedling for a longer time in the stimulated position. By exposing for successively longer and longer periods until a reaction results, the presentation time can be determined.

Reaction Time—12. Having determined the presentation time, the reaction time can be measured, this being the interval between the first exposure to the stimulus and the commencement of the reaction.

Perception Time—13. Having determined the presentation time needed for a Bean seedling to react to gravity when placed in a horizontal position, the perception time can be found by exposing seedlings for successive fractions of the presentation time, placing them in an unstimulated position, e.g. vertical or on a revolving klinostat during the intervals between the exposures; e.g. assuming the presentation time to be twenty minutes, twenty periods of one minute, ten periods of two minutes, five periods of four minutes, and so on, can be successively tried. The *shortest* period which is effective when used as a fraction of the total time required is the perception time. Using the same method, the maximum duration of the interval between the exposures can be determined.

Hydrotropism—14. The reaction to moisture is especially strong in the case of roots and will overcome that of gravity. It can be demonstrated by germinating seeds of e.g. Mustard in a small tray, the bottom of which is perforated or made of coarse muslin. Keep the air in the tray moist by means of damp cotton-wool and place it over a dry chamber, the bottom of which is covered with a layer of calcium chloride. It will be found that the roots of the seedlings, instead of growing straight downwards, will turn up and grow along the bottom of the moist sieve. *N.B.*—This reaction is evidently a special case of chemotropism.

Contact Irritability—15. Reaction to contact may be observed by gently touching the base of the stamens of Barberry or Corn-flower, and noting the movement which ensues. In the case of tendrils, select a long, straight tendril of Bryony, Vegetable Marrow or Passion Flower. Gently rub the surface of the tendril near the tip with a piece of stick or a pencil, and note and record the curvature which follows.

Circumnutation—16. Take a young, actively-growing plant in a pot and stand it in a box in such a position that it will be equally illuminated on all sides. Attach a fine, straight pointer to the growing region of the stem by means of gelatine, so that it is in line with the stem and projects above it. Cover the box with a sheet of glass and make hourly observation of the exact position of the pointer. The course of the movement can be recorded on the surface of the glass by means of dots made with India ink or with a pencil suitable for writing on glass. The dots can be joined up and a tracing of the whole made when the record is complete. In recording each position the eye must be in line with the top and base of the pointer.

Etiolation—17. Sow seeds of Broad Bean, Mustard, Wheat, Barley, or Oats in a number of pots. Place pots of each in the light and others in a dark cupboard, and keep in the same conditions after germination. Make careful records of the two lots of seedlings for some weeks and compare the etiolation phenomena shown by the different seedlings. The pots should be adequately watered at sowing. Those germinated in a room or in a dark cupboard should be stood in saucers and can easily be watered from below without disturbing the seeds by placing water in the saucers.

CHAPTER XIV

ECOLOGY AND PLANT GEOGRAPHY

Introductory—It is a matter of general observation that
certain species of wild plants are commonly found growing
together, and associations of this kind occur with such
frequency that they are not likely to be merely accidental.
Further, if we reflect on the meaning of such common words
as *heath, down, meadow* and *fen*, it becomes apparent that we
are accustomed to associate certain kinds of plants with
definite, more or less easily recognisable conditions of soil and
climate. Coincidences of this kind are so usual that they have
become fixed in names such as those mentioned, and we do
not need to be botanists in order to recognise that wild plants
group themselves naturally so as to give rise to distinct types
of vegetation which are often characteristic of particular
localities.

When we walk across a heath, for instance, we expect to
find plants such as Gorse, Heather, Sheep's Sorrel and
Tormentil. We do not expect to find there oak woods car-
peted with Bluebells or Primroses; we do not expect, indeed,
to find woods at all in such a situation, and we know that
if trees are present they will almost certainly be either Scotch
Firs or Birches. The localities where heaths occur are freely
exposed to wind and sun, and such areas are most common
in districts of relatively low rainfall; they are, in consequence,
more characteristic of the south-eastern part of Great
Britain than of the north-western part. The soil is usually

sandy or gravelly, with a few inches of brown peaty mould
on the surface, and is notably deficient in chalk. It is not a
soil of a productive or fertile kind; and indeed the very existence
of a piece of waste land such as a heath is evidence that, for
one reason or another, it is not profitable to cultivate it.

Now it is clearly the business of the botanist to apply his
knowledge of the individual plant and its manner of life to
the study of plants as they occur in nature, and to investigate
the causes which underlie their distribution. The solution
of such problems will severely test his ability and botanical
training. He must have a sound and extensive knowledge of
Systematic Botany in order to recognise and record the
different plant species present, and will then need all the
knowledge and experience gained in the laboratory before
he can hope to solve, or even to state clearly, the problems
which confront him in the field. Nor is it a matter solely of
botanical interest that such investigations should be under-
taken by botanists, since they may lead, directly or indirectly, to results of great practical value.

The problems which confront the student of natural
vegetation are obviously akin to those which the farmer and
gardener are unceasingly, although often unconsciously,
trying to solve. On the farm or in the garden we are dealing
with groups of plants which have been sown or planted,—that
is to say, with artificial vegetation as compared with wild
or natural vegetation. It is the business of the farmer and the
gardener to so modify and control the conditions under which
his artificially produced vegetation is growing, that each
individual plant may reach its greatest development and
so contribute to the largest possible crop. It is one of the
duties of the modern botanist to show that the knowledge of
plants which he has acquired by observation and experiment
is not solely of intellectual interest, but that it can be applied
to practical problems such as increasing the food-supply of
the world, improving the amount and quality of timber,
and the production of a larger yield and better quality of
vegetable materials important to industry, such as cotton,
linen, jute and rubber.

The branch of Botany which is specially concerned with the
study of vegetation,—with the study of plants " at home "
or, as it ·is described in scientific language, with the study of

plants in relation to their *habitat,*—is known as *Plant Ecology,* the word " ecology " being derived from the Greek word οἶκος, meaning a house or home.

The first business of the plant ecologist was to learn to recognise the different types of vegetation. It was then necessary to name them and to devise a classification which would have the approval of plant ecologists everywhere, since, in every branch of work, it is of the first importance that fellow-workers should be able to exchange views, and feel sure that when they use the same word they mean exactly the same thing.

Botanists in this country, in various parts of Europe and in America, have already made extensive studies of vegetation from the ecological point of view, and much progress has also been made towards devising a satisfactory system of classification. This part of the subject need not detain us long, but it may be helpful at this stage to illustrate what has been said in order that the student may understand the point of view adopted by the plant ecologist who attempts to distinguish the different groups of plants which compose the vegetation of a district. It is convenient for this purpose to select for description a place where the vegetation has not been altered to any extent by cultivation.

The Plant Society—Let us take, for example, a Beech wood on the slopes of the South Downs. If we visit such a wood in spring, just before the trees are in leaf, we will probably notice large patches of a common plant, the Dog's Mercury (*Mercurialis perennis*), under the trees, and we may note also that even at this early season the plants are in full flower and vigorous growth (Plate I.). In this particular situation, Dog's Mercury can evidently grow vigorously and spread rapidly to the exclusion of other plants; it is, in fact, a good example of what the ecologist calls a *plant society.*

This is the simplest kind of vegetation group or *unit* which we can imagine, and it evidently originates as a result of the multiplication of some individual plant species so as to give rise locally to a continuous covering of vegetation. Plant communities do not occur everywhere and are often due to accidental causes. They are found when a species which is able to multiply rapidly, either by means of seed or in some other manner, is so favoured by local conditions

14

that it can monopolise the ground to the exclusion of all competitors.

In the case of Dog's Mercury, the plant is a perennial, spreading rapidly by means of underground stolons or rhizomes. It competes successfully with other plants and so forms extensive patches or plant communities in the Beech wood because of its habit of starting growth in the early spring, before development is handicapped by the dense shade cast by the trees as soon as the leaves unfold. Incidentally, Dog's Mercury and the Beech trees beneath which it so often grows are a good example of what is called a *complementary plant society*. The two plants thrive together, because one passes through its period of active growth and bears flowers and fruits while the other is still in a leafless and comparatively dormant condition. The food manufactured by the leaves of the Dog's Mercury in spring is collected in the underground parts, and these serve as a storehouse upon which the plant can draw for its growth in the following year.

The Plant Association—Continuing our observations on the Beech woods, so characteristic of parts of the North and South Downs, it will soon be noted that trees of other species occur amongst the Beeches, more especially at the edges of the wood. In the language of the plant ecologist, Beech is the *dominant* species, associated with which are *subordinate* species, notably the Yew (*Taxus baccata*), and the Whitebeam (*Pyrus aria*), with shrubs such as Box (*Buxus sempervivum*), Privet (*Ligustrum vulgare*), Spindle Tree (*Euonymus europæus*), and Guelder Rose (*Viburnum Lantana*).

The relative importance of these subordinate species varies with the locality; Yew and Box, for instance, may become locally dominant, and place-names such as Box Hill and Boxford illustrate the way in which such local characters of the vegetation may be put on record.

In these Beech woods the undergrowth is scanty and consists of only a few species, some of which are specially characteristic of this situation and are rarely met with elsewhere. Thus, in addition to Dog's Mercury, a plant common in shady places, we notice Wood Sanicle (*Sanicula europæa*), Enchanter's Nightshade (*Circæa lutetiana*), Dog Violets (*V. Riviniana* and *V. sylvestris*), locally certain orchids e.g.

the White Helleborine (*Cephalanthera Damasonium*), and the Bird's Nest Orchid (*Neottia nidus-avis*); more rarely still the Yellow Bird's Nest (*Hypopitys Monotropa*). The two plants last mentioned are curious and characteristic members of the ground vegetation beneath Beech trees. Both are total saprophytes without green leaves, and appear above ground only during the flowering and fruiting periods.

Comparison of one Beech wood with others in similar situations shows that the same plant species tend to recur, although their relative abundance varies with changes in local conditions. This indicates that the composition of the vegetation is not accidental or fortuitous, but, given the natural conditions provided by the slopes of a chalk down, certain species prosper more than do others, and, eventually, as in the case of a heath, we find a characteristic assemblage of plants associated with special conditions of soil and climate. We recognise, in fact, what the plant ecologist calls a *plant association*, which takes us another stage on the way towards the analysis and recognition of different types of natural vegetation. Such *plant associations* may, or may not, include *plant societies*. Since they are associated with definite conditions which can be described and analysed, it is evidently of the first importance that the student of vegetation should be able to distinguish them.

Extending our observations to the open down near the Beech woods, we become aware that the whole of this elevated downland country provides conditions of a rather exceptionally uniform kind, and a general survey of the vegetation shows that it includes more than one type of plant association. In addition to the Beech wood association already studied and the more uncommon type of wood in which Yew is the dominant tree, there is the characteristic short springy turf of the open down, composed of a few species of grasses, associated with plants such as the Kidney Vetch (*Anthyllis vulneraria*), the Rock Rose (*Helianthemum vulgare*), the Burnet (*Poterium Sanguisorba*), the Horseshoe Vetch (*Hippocrepis comosa*) and a number of others.

We find, in fact, that the vegetation of a chalk down can be classified without much difficulty into one or two *woodland associations*, a characteristic *grassland association*, with, in places, an association intermediate between them, possibly due to the degeneration of woodland from various causes, a *scrub association* as it is called. Dwarf bushes of Yew, Juniper and Hawthorn are often locally abundant on the Downs, forming a plant association of this kind.

The Plant Formation—Our vegetation survey has thus led us to the conception of a number of plant associations making up the vegetation of a region with characteristic natural features,—to the conception, in fact, of what is known to the ecologist as a *plant formation*. This vegetation unit cannot be so easily defined as can the plant association, and its recognition requires more experience. It may be thought of as the whole of the vegetation occupying a situation or a habitat in which the natural conditions are fairly constant and uniform.

Now the natural conditions existing on a chalk down are determined mainly by geological considerations. Chalk is a rock which does not wear away or weather rapidly and it therefore gives rise to country of some altitude. The rounded contours of the Downs of south-east England and of the Yorkshire and Lincolnshire Wolds are characteristic, and reach in places a height of 900 feet or even 1,000 feet above the sea. The thin, friable soil accumulates but slowly on the steep slopes and contains a very high proportion of carbonate of lime: such a soil is known as a *calcareous soil*, and when cultivated has special characters which must be taken into account by the farmer if he wishes to grow satisfactory crops.

In Great Britain calcareous soils are found on limestone, chalk and certain calcareous sandstones and clays, where these rocks come to the surface. Extensive areas occur on the Older Limestones, and on the Chalk which is, chemically, relatively pure and very soft limestone. Of the former, the most important regions are the Mountain Limestone districts in the Pennines and Wales, the Mendip Hills and parts of Ireland, the Magnesian Limestone forming hilly country in Derbyshire and other parts of northern England, and the Devonian Limestone in various parts of Devonshire. Similar soils are also found on the Oölitic Limestones of the Cotswold hills and elsewhere. The Chalk comes to the surface in southern, eastern, and north-eastern England, where it gives rise to the characteristic uplands known as *Downs* in the South and *Wolds* in the North.

Owing to the fact that the Older Limestones occur chiefly in the north and west where the rainfall is greater, their vegetation differs in several respects from that on the Chalk, although many lime-loving species are common to both. In the north of England, for instance, Ash replaces Beech as the dominant tree of the chief woodland association and with it are associated Wych Elm, Hawthorn, and other subordinate species. A grassland association is found which corresponds closely with the short turf of the Chalk Downs, the dominant species of grass in both cases being the Sheep's Fescue (*Festuca ovina*). In parts of Yorkshire, limestone pasture may be largely composed of Moor-grass (*Sesleria cærulea*), a grass peculiar to the calcareous soils of the district.

Grouping together the plant associations found in the chalk and limestone districts of Great Britain, we can form a mental picture of a plant formation, subdivided, if we wish, into sub-formations, characteristic respectively of the Chalk and of the Older Limestones, and call it the *plant formation of calcareous soils*.

" *Calcareous Heath* " and " *Limestone Heath* "—On the crests of the Chalk Downs a much deeper soil is often found and we may have to dig for several feet before the surface of the chalk rock is reached. Carbonate of lime is readily washed out of the soil by rain, and these deposits on the crests or level parts of the Downs may be formed by the " leaching " of calcareous soil in this way, or by the presence of superficial deposits of different origin deficient in calcium carbonate. They compare sharply with the thin soils on the slopes owing to their poverty in calcareous material.

With this alteration in the nature of the soil is usually associated a sharp contrast in the vegetation. We find mixed with, or replacing, the plants of the grassy slopes, a number of other species, such as Heather (*Erica cinerea*), Ling (*Calluna vulgaris*), Gorse (*Ulex europæa*), and Tormentil (*Potentilla erecta*)—in short, many of the plants which make up the chief plant association of heaths. This provides a remarkable illustration of how a change in the external conditions, in this case the absence of one constituent from the soil, may determine the nature of the vegetation.

Most of the plants found on heaths refuse to grow on

soils containing appreciable amounts of carbonate of lime.
Many of the plants of the Downs grow best on chalky soils,
to a few of them a high proportion of this particular soil-
constituent is indispensable. In the case described, the lime-
loving or *calcicole* plants of the chalky slopes cannot compete
successfully with lime-hating or *calcifuge* plants, and the
latter thus give rise to the anomalous kind of vegetation
known as " calcareous heath." Similar causes give rise to
" limestone heath " in the limestone districts, in the higher
and more exposed parts of which thick deposits of peat may
accumulate.

The ability of lime-loving species to survive on calcareous
or limestone heaths may depend on the depth to which they
can send down their roots. Deep-rooting plants such as
Burnet and Kidney Vetch may be found growing side by side
with Heather or Ling, because the roots of the former can
penetrate into the chalky débris below the soil. We have here
another good example of a *complementary society*, the plants
in this case flourishing together because they make demands
upon different layers of soil.

In a similar way we can study the vegetation composing the
plant formation of sand dunes, the *plant formation of heaths*,
and so on.

For a fuller account of the different plant formations into
which the vegetation of the British Isles has been distinguished
reference must be made to a special textbook.* It must be
understood, however, that the vegetation of a comparatively
small area of this country has as yet been mapped or studied
in detail, while experimental work on the vegetation problems
which everywhere present themselves has hardly been
begun. A very rich field of exploration and research lies
before the student of vegetation, one from which he may
hope to gather results of more than theoretical interest.

Following the method described in the preceding pages,
the student will best gain experience by undertaking a
detailed study of a small area of natural vegetation and
recording it in the form of a vegetation map, having first
made himself familiar with the local conditions of soil and
climate.

* See " Types of British Vegetation." Edited by A. G. Tansley.
Cambridge University Press. 1911.

Vegetation Changes—We have hitherto discussed plant groups, or *vegetation units*, as if they were stable and permanent things. The critical student will at once object that this is not the case, but that the composition of any given piece of vegetation may and does change from year to year. Owing to some accidental cause,—a very severe season, or destruction of vegetation by fire,—the proportion of the different kinds of plants present may be altered so as to change greatly the nature of the vegetation as a whole.

Changes in vegetation depend upon the operation of causes which may be classed under two heads: (1) Causes responsible for the appearance of new plant species, and (2) causes which determine the ability to survive or the relative abundance of species already present. Causes included under (1) are often accidental; thus, the effectiveness or otherwise of the means whereby seeds and fruits are distributed, or the proximity of species which spread rapidly by vegetative means, will evidently determine the appearance of new plants in the first instance.

The subsequent course of events depends upon the nature of the vegetation; if the ground is untenanted, or there are bare patches, the chances for seedlings to find a foothold will evidently be better than if the ground is closely covered with plants. It is important to distinguish between an *open formation*, in which the ground is partly bare, and a *closed formation*, in which the surface is completely colonised with plants.

In a closed formation, such as the short turf of the Chalk Downs, it is difficult for any new plants to gain entrance, and the chance of doing so depends often on the presence of accidentally bared areas, such as mole-hills or ant-hills.

If a piece of ground is artificially bared and allowed to acquire a new covering of vegetation by natural means, the first representatives of the new population will be seedlings of plants which possess effective arrangements for seed dispersal, or those which can spread rapidly by vegetative methods. As time goes on and competition for room becomes keener, those plants are most likely to persist which are favoured by the local conditions of soil and climate. These factors will eventually determine the proportion of dominant and subordinate species present as a result of *natural selection*,

—a process which inevitably involves the disappearance of species, " unfit," in the sense that they are unsuited to the local conditions.

Thought of in this way, any piece of natural vegetation consists of a number of plant species *selected* from those available by reason of their suitability to the local conditions.

Each of the vegetation units described may be considered as such a case; and in so far as each has been formed slowly by the encouragement of species well suited to the habitat and the elimination of others less favoured, it does represent a more or less stable and permanent arrangement. Not all vegetation units have reached this condition of comparative stability; some of them may be still in the experimental stage, so to speak, and if carefully observed over a period of years will yield evidence that this is the case. Furthermore, local conditions alter and a slight disturbance may be sufficient to produce relatively great changes in the vegetation.

In some parts of Great Britain there are extensive areas of moorland on which are found deep deposits of peat. Owing to special conditions of soil and climate this peat has been built up by the slow accumulation of many successive layers of partly decayed plant remains. The plants which form the vegetation of the moor to-day,—Heather, Ling, Bearberry, Bilberry on the drier parts; Cotton-Grass, Sedges, *Sphagnum* in the boggy parts,—are growing upon a mass of peat composed of the remains of plants which in past ages formed the vegetation of the same place.

By a careful examination of peat, it is often possible to reconstruct the earlier vegetation which gave rise to it; fruits, seeds, roots, stems and leaves are often so well preserved that they can be identified without difficulty. An investigation of this kind was carried out some years ago on various moors in the north of England and Scotland. Holes were dug in the peat and samples from different depths carefully examined for fruits, seeds, or other recognisable remains. In places, as many as nine distinct layers of peat were found, built up one above the other, each consisting of the remains of plants which had previously composed the vegetation. Each of these layers of peat represents a very long period of time, many times longer than the lifetime of an individual man, and collectively they form a very remarkable record of the history of the vegetation of that place during a long period of recent geological time.

From the plants found in each layer, it was apparent that great fluctuations of climate had taken place, and that one kind of plant association after another had been formed and had given way under the stress of altering conditions of climate.

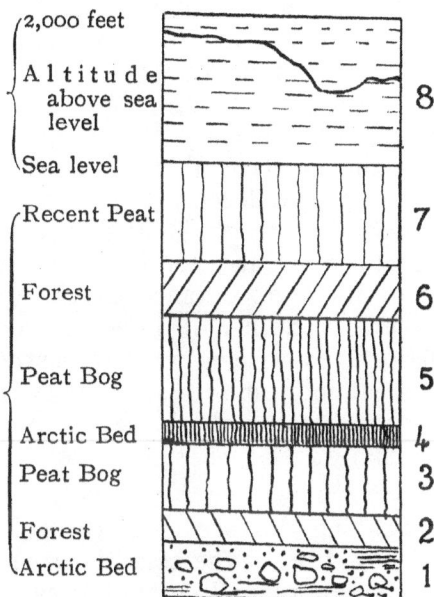

FIG. 33—DIAGRAM TO ILLUSTRATE THE SEQUENCE OF VEGETA-
TION IN A DEPOSIT OF PEAT NEAR TWEEDSMUIR

1, Remains of arctic plants. 2, Remains of Birch and Willow. 3, Peat-bog plants. 4, Remains of Arctic plants. 5, Remains of peat-bog plants. 6, Remains of Birch. 7, Recent peat with remains of Ling, etc. 8, Small-scale section of Tweedsmuir peat area showing present contour and elevation (After Lewis)

In some of the deeper layers were found remains of plants growing now only in very cold or arctic regions; others gave evidence of a woodland association composed of trees, such as Alder and Birch; yet others yielded remains of plants such as are found in our peat bogs to-day.

This is a record of vegetation extending over vast periods of time, but we have reason to believe that changes of a corresponding kind are everywhere taking place, sometimes with comparative rapidity, sometimes so slowly as to be imperceptible. In the Peak district of Derbyshire, for example, where the vegetation has been carefully studied and mapped, marked changes have occurred during the past 60 or 70 years. Streams have cut their way back in times of flood, so causing changes in the drainage and consequently in the water-supply of plants growing in the vicinity.*

Facts such as these teach us to think of the plant formation or plant association as possessing an individuality of its own, comparable with that of an individual plant or other organism, which comes into being, goes through a period of active growth and development, reaches a condition of maturity when growth and change seem almost to have ceased, and then undergoes changes of a degenerative kind and ultimately dies. In a corresponding way, changes may take place in any piece of vegetation that alter its character, and may eventually lead to its disappearance as a vegetation unit and its replacement by another of different type. We must not press the analogy too closely, but, to the student of plant ecology, it is of great interest to search for evidence of such changes, and to try to discover whether a given piece of vegetation is in a condition of relative stability or not at the time of observation.

The Habitat—Leaving this aspect of the subject, we will briefly consider the nature of the conditions which make up what has been spoken of as the *habitat* of a plant.

The term *habitat* includes all the natural conditions which exist where the plant is growing. If we think of these conditions separately, we can call each of them a *factor*, and it is necessary to consider only such factors as affect the plant directly or indirectly,—*effective factors* as they are called. Such will evidently include *physical* factors related to climate and soil, and *biotic factors* due to the action of living organisms.

For convenience of reference, the more important of these effective factors are summarised in tabular form (p. 219).

It is convenient to consider physical factors in two groups, (*a*) those relating to climate, and (*b*) those relating to the soil.

* See " Vegetation of the Peak District." By C. E. Moss. Cambridge University Press. 1913.

TABLE SHOWING THE CHIEF FACTORS IN THE
ENVIRONMENT OR HABITAT OF A PLANT

	(a) *Climatic:* temperature; rainfall; humidity; altitude; wind; light; aspect; exposure; slope of ground.
A. *PHYSICAL FACTORS.*	(b) *Edaphic* (soil): the chemical nature of the soil, in so far as it affects the food supply of the plant; the physical character of the soil, in relation to the water supply; *indirect* biological effects due to the presence in the soil of living organisms: earthworms, microscopic forms of animal life, Fungi and Bacteria.
B. *BIOTIC FACTORS.*	Interference by man; grazing animals of all kinds; attacks of parasites; the influence of one plant upon another; *direct* effects due to Fungi and Bacteria.

Among climatic factors are, temperature, rainfall, amount of moisture in the air (humidity), height above the sea (altitude), wind and light, incidental to which are aspect, exposure, and slope of the ground. Such factors are very important, especially those relating to temperature and rainfall. Minor conditions of climate are important also, and the effect of all such factors must be carefully studied by the plant ecologist.

Edaphic factors include those determined by the chemical nature of the soil as well as by its physical characters, the latter being directly related to the water-supply of plants growing on it. The presence of water from infiltration of rivers and the indirect effects due to the presence of living organisms also comes under this head.

Physical Factors—I. The Vegetation of the World— Broadly, the vegetation of the earth's surface may be classed as *woodland, grassland,* or *desert.* In the temperate and tropical zones the type of plant covering is primarily determined by the amount and distribution of the rainfall, the humidity of the air, and the prevalence of winds in so far as these have a desiccating influence. The constitution of the vegetation floristically, i.e. as regards representative species of the flora, is chiefly controlled by temperature.

While the general features are determined by climate, the local details may be dependent on edaphic factors,—especially the water content of soil. For example, in a dry region with small rainfall, characterised generally by sparse vegetation, there may be belts of woodland along river banks owing to the infiltration of water into the surrounding soil, stagnant water may determine the existence of a swamp, and so on.

Woodland and grassland stand opposed to one another, the limits of each being determined by the climates which now prevail. Any alteration of climate, whether from natural or artificial causes, may lead to replacement of one type by the other. Thus, the clearing of forests by man, by lowering the humidity of the air, may so alter the climate that grassland replaces woodland in neighbouring districts. Success in the struggle between woodland and grassland is attained by the type with which the climatic conditions best accord.

Woodland Climate—The majority of trees demand plenty of water because of their big root development and the great transpiring surface they present to the air. They can withstand long seasons of drought and heat because of their deep root systems (e.g. in the Mediterranean region). The trees in districts with a dry vegetative season are dependent on water-supplies collected during the wet season and stored in the deeper parts of the soil. So long as a supply of water is assured to the deep-growing roots it does not matter at what season the supply is renewed, or whether it is derived from frequent small supplies of rain, or from heavy rainfall confined to a short rainy season. On the other hand, in higher latitudes, a climate with a dry winter, especially when accompanied by drying winds, is hostile to woodland, since the trees cannot replace from the frozen soil the water lost in transpiration. The habit of losing their leaves in winter shown by many trees in temperate climates is related to this danger of desiccation in winter.

The most luxuriant type of woodland is found in the region of tropical rain forest where high temperature, great light intensity and constant humidity of the air encourage vigorous growth. The high light intensity often results in the development of dense undergrowth which can only be penetrated by cutting a way through; in other cases, however, under-growth is practically absent, since the denseness of the overhead foliage is such that only enough light penetrates to allow the growth of a few Ferns. The struggle for light is evidenced by the large number of " climbers " and " epiphytes " (plants attached to other plants and not to the soil). Such rain forests are evergreen in that the leaves of the trees are lost at irregular times, so that while some trees or branches of some trees may be without leaves, the majority bear foliage. In tropical forests with a pronounced dry season or temperate forests with a dry winter, the majority of trees lose their leaves at the same time and for this resting season the forest is, for the most part, leafless.

In the colder temperate climates coniferous forests are common, which are evergreen for the reason that the foliage leaves persist for more than one season, and the trees therefore are never completely devoid of leaves as are broad-leaved deciduous trees.

Grassland Climate—As compared with trees, grasses are shallow-rooted plants, and it is consequently only the moisture available in the superficial layers of the soil which is im-portant for them. Grassland, therefore, requires frequent even if slight showers *during the growing season* so that the superficial layers of soil may be kept moist, and demands a moderate temperature during the same period. Grasses grow close to the ground where the air is both damper and less disturbed in windy weather; moreover in many grasses, the stem grows underground as a rhizome. Thus, dryness of the air, or wind, even when the soil is cold, does little harm to grassland *during the resting period*.

Extensive grassland is found in the North American *prairie* districts (between the Mississippi and the Rocky Mountains) and in the *steppes* of southern Russia. The development of grassland in the prairies is due to the dry winter, with frequent winds and severe frost, the moist early summer and moderate rainfall,—insufficient to maintain good forest growth. In the Atlantic district to the east of the Mississippi, where the rainfall is much higher and occurs partly during the winter, forest is developed. The Russian steppes likewise possess a grassland climate. The rainfall is only moderate and falls chiefly in spring and early summer;

this is followed by drought in late summer and autumn (at the end of the vegetative season for grasses), while the winters are cold with strong, dry east winds.

" A woodland climate leads to victory on the part of " woodland, a grassland climate to victory on the part of " grassland. In transitional climates "—such as that of the British Isles—" edaphic influences decide the victory. Strong " deviations from woodland or grassland climate produce " desert."*

Deserts—The most marked characteristic of desert regions is the low rainfall, which never exceeds 12 inches and is usually very much less. The few plants which manage to exist under desert conditions fall into two ecological groups, (1) those which depend directly on rain, (2) those which depend on the presence of subterranean water.

The former class includes annuals and perennials. The annuals germinate at the commencement of the rain and at its cessation ripen their seeds and die. It is somewhat remarkable that, apart from their very rapid development and the shortness of their life, they show no special adaptation to the climate. The stems and leaves are often delicate and herbaceous, the roots fine and the flowers large. The perennials comprise plants with underground resting organs. They appear above the ground after the rain, but die away again with the resumption of the drought.

The plants dependent on subterranean water are characterised by the immense length of the root system, which must reach down many feet before water is met with. The leaves of such plants, usually, but not invariably, are constructed so as to lose water only with slowness.

Sharply contrasting with the general desert, with its scanty vegetation, are the *oases* or depressions containing springs, which may support a luxuriant growth of plants and are invariably under cultivation.

Arctic Climate—The chief characteristics of the arctic climate are the long, cold winter and the short, cool summer with continuous daylight. The actual cold is, in general, no more severe than at many points in the temperate zone,—more destructive to plant life is the extreme dryness of the air and the severe gales often prevailing. Nearly everywhere,

* Schimper, " Plant Geography," p. 174. Eng. Edn., 1903. Oxford.

even in summer, the temperature of the air is low, however, owing to the presence of large masses of ice. Exposure to direct sunlight, or *aspect*, is therefore much more important in raising the temperature of the soil than in lower latitudes. The frequent occurrence of wet fogs such as is characteristic of Spitzbergen in summer must consequently have a great influence on vegetation.

Owing to the prevailing low temperature of the soil, absorption by the roots is retarded and plants exhibit effects such as stunted growth, promotion of flower formation, etc., similar to those induced by dryness of soil. The continuous light during summer, while favouring assimilation and the formation of pigments, also exerts a retarding action on growth. The shortness of the warm season necessitates a rapid vegetative period and involves much danger to the ripening seed.

By way of summary we may say that the vegetation is stunted and chiefly developed on slopes exposed to direct sunshine, while the plants develop their flowers, which are usually brilliantly coloured and in considerable abundance, quickly, as otherwise the seeds would not have time to ripen before the return of winter. It is to be noted that arctic plants possess no structural features which can be recognised as protecting them from cold, although they usually show modifications to check transpiration and guard against desiccation, which is threatened by the difficulty of absorption from the cold soil. In less favourable regions, which constitute a " cold desert " or *tundra*, if vegetation of any kind occurs, it is practically limited to mosses and lichens.

Mountain Climate—The changes in climate which take place as we pass from a high to a low elevation may be compared in many ways with the changes which occur as we pass from lower to higher altitudes. Increase of altitude is accompanied by a corresponding lower atmospheric pressure and this results in several secondary consequences. In the first place, the sun's rays are less absorbed by more rarefied (and dryer) air, so that with a higher altitude there is a lower temperature. Even if warm air from the lower layers rises, it expands as the pressure diminishes, and so automatically becomes cooled. This effect is intensified by the presence of masses of snow or ice on the higher parts of the mountains.

On the other hand, since less heat is absorbed in passing through the more rarefied air, the intensity of direct sunlight is greater at high altitudes,—especially in regard to the blue end of the spectrum. Situations exposed to direct sunlight may therefore attain quite high temperatures, although the loss of heat at night is correspondingly greater, since radiation occurs more readily through the rarefied air. There will thus be great extremes between day and night temperatures. Other characteristics of the climates of high mountains, such as the Alps, are great dryness of the air in winter and the fact that the air is nearly always in motion, causing great intensity of evaporation.

In respect therefore to the great importance of direct sunlight, the intense evaporation and low temperature, the greater intensity of illumination (which appears to exert effects analogous to those produced by continuous illumination), and the short summer growing season, a mountain climate approximates to that found in arctic regions, and alpine and arctic plants show many common features, e.g. profuse flowering, brilliant flowers, stunted habit, short vegetative season, etc., and structural modifications of the foliage leaves which protect them from excessive loss of water during transpiration.

Physical Factors—II. The Vegetation of the British Isles— In a small country like Great Britain where the climate is relatively uniform, differences in the soil predominate in their effect upon vegetation. Although climatic factors must always be taken into account, their effects are often overshadowed by those depending on differences of soil,—*edaphic factors*, as the plant ecologist calls them.

This fact is expressed in the names given to many of the British plant formations,—e.g. *the plant formation of siliceous soils, the plant formation of clays and loams, the plant formations of calcareous soils* (p. 213), etc. An understanding of the relation of plants to the soil is of great and fundamental importance to the botanist, and is considered more fully in Chapter XV.

The influence of climatic factors on natural vegetation can be studied by comparing the western and north-western parts of the British Isles with the eastern and south-eastern parts of the country. In the former there is a much heavier rainfall, especially in the mountainous districts, and with this is correlated differences in the vegetation.

Thus, there is a greater development of woodland in the west and north-west and in these woods a greater abundance of Ferns. This tendency is partly counteracted, however, by the greater exposure to prevailing westerly winds and the more mountainous character of the country, so that moorland with a characteristic assemblage of plants is prevalent in the less sheltered areas.

The influence of climate can also be studied in this country in relation to altitude and many of the effects already described (p. 223) may be noted when ascending a high mountain.

Biotic Factors—Biotic factors may exert marked effects upon vegetation, and their importance in this respect has been often overlooked. Among such factors are included those due to interference by man, those due to grazing animals, and those depending upon the interaction of a plant with its neighbours or with micro-organisms, such as Fungi and Bacteria. Full of interest as are these matters to the student of plants, the space at our disposal does not permit us to study them in detail. Some of them are touched upon again in the chapter on Soil and elsewhere (pp. 117, 250).

One or two illustrations selected from the not very numerous cases in which the effects of biotic factors on vegetation have been carefully studied, may serve to suggest starting points for fresh observations on the subject.

An interesting case of a marked change in vegetation resulting from unconscious interference by man has been recorded recently from one of the eastern counties of England. The district is one of low rainfall, with a soil so loose and sandy that plants suffered severely from drought in dry weather. As a practical consequence, it was found impossible to make a golf links in the neighbourhood, owing to the difficulty of maintaining a grassy turf. It so happened that this district was selected for prolonged army manœuvres, in the course of which the ground was consolidated by much trampling and marching. This improved the water-holding power of the soil, and consequently the water-supply to the upper layers, as a result of which there was formed and maintained in the following season a continuous covering of

15

turf, composed of the rather wiry, fine grasses characteristic of heaths.

This serves as an example, not only of a marked change in vegetation indirectly due to human agency, but also as one depending directly on an alteration in soil conditions. The nature of the relation between the physical condition of the soil and the water-supply of the plant is considered later (p. 238).

Effects produced by grazing animals are more obvious and familiar; one may instance the remarkable changes in vegetation which can take place by reason of a preference by rabbits for one plant rather than another. A recent study of an area of heath where rabbits are plentiful, and where vegetation changes of a marked kind are taking place has shown that owing to the severe attack made by these animals on certain heath plants, especially on Ling and Heather, these plants succumbed and were replaced by Grasses, which can withstand an almost indefinite amount of grazing. In some seasons plants may be attacked by caterpillars or by various parasites, and may suffer severely in consequence. Damage of this kind is, however, usually of a temporary character, and is without permanent effect on the vegetation (Plate VI., A).

To pass on to effects depending on the interaction of one plant with another. The Bracken (*Pteris aquilina*) is an extremely successful plant which, given suitable conditions of soil and climate, dominates the vegetation of large areas. It competes successfully with most plants, not only by reason of its vigorous growth, comparative immunity from attack by grazing animals and effective means of vegetative spread, but also on account of its effect upon neighbouring plants, which may eliminate competition by clearing the ground of all competitors. Bracken casts a very dense shade, not only in the growing season, but by the persistent accumulation of dead fronds in winter. It is therefore practically impossible for any herbaceous green plant to grow beneath it, and since vigorous fronds may reach a height of 8 feet, it can even smother and kill small bushes of Hawthorn and Gorse* (Plate VI., B).

Among practical gardeners the view has long been held that certain plants will not grow side by side because one exercises a poisonous effect upon another. Until recently

* See Farrow, *Journal of Ecology*, vols. v. and vi.

PLATE VI

A

B

THE INFLUENCE OF BIOTIC FACTORS ON VEGETATION

A, the effect of rabbits on Heather. Three closely-grazed hummocks of Heather are seen in the middle of the bare area. By attacking the Heather on the edge of this area, the rabbits probably lead to its extension.

B, the effect of Bracken on other plants. The rhizomes of the Bracken are advancing towards the middle of the Gorse bush from the left of the picture. The middle of the bush, covered by the dead fronds, is dead : the portions on either side are living and healthy.

it has been the fashion for botanists to smile at these statements, but there is now a certain amount of experimental evidence that something of the kind may actually happen. It is known, for example, that young fruit-trees do not thrive in an orchard unless the turf is removed for some distance round the base of the trees. This has been attributed to competition on the part of the grass for water and plant food from the soil, but experiments have shown that if plants are grown in pots on which are placed perforated trays of soil with growing plants, so that all water supplied to the experimental plants passes through the soil in which the plants above them are growing, the former become sickly and cease to grow.

It is not possible at present to offer an explanation of these facts, but they allow us to conclude that one growing plant may exert an effect upon another, although the way in which it does so is at present unknown.

When we turn to effects produced by micro-organisms we are confronted by a vast array of facts, many of them of extreme interest, of which we can here give only the briefest outline. It should be noted that our classification of factors is not entirely logical, since among biotic factors must be reckoned those depending on the action of micro-organisms living in the soil.

In dealing with biotic factors due to the action of micro-organisms, we have, as a matter of convenience, distinguished between the cases in which a direct relation is known to exist between plant and micro-organism, and those effects of an indirect kind due to the action of soil organisms. The latter are not treated here as separate *biotic factors*, but will be considered later under the heading of *biological soil factors* (p. 250).

Of the former, two types require special mention. *First*, there are the parasites which attack green plants and are the chief cause of the numerous diseases which afflict them. Among cultivated plants, such diseases cause the loss of thousands of tons of valuable foodstuffs every year and the problem of dealing with them is one of the most serious with which the farmer and the gardener has to cope. Among such diseases are the "rusts" which affect the grasses and cereals, the "mildews," which in some seasons do such widespread damage in garden and hedgerow, the dreaded " potato disease," so difficult to combat in damp warm weather, and a host of others.

A majority of diseases of the Higher Plants are caused by Fungi; others, in common with diseases of human beings such as pneumonia and diphtheria, are due to parasitic bacteria which enter the tissues of the plant, multiply there and eventually give rise to " rots " of various kinds by destruction of the infected parts.

Others again are caused by attacks of lowly animal parasites, such as species of eelworm, which enter the roots from the soil and produce a curiously distorted condition of the diseased parts.

In the *second* place, there are the cases, fewer in number, but of unusual interest, in which Fungi or Bacteria enter the tissues of a plant, but having done so do not behave as typical parasites. It is as if the plant and its invader, instead of assuming the rôles of host and parasite, had struck a bargain and formed a kind of mutual benefit society from which both partners can derive advantage.

In some plants this appears to be actually the case; in others it would not surprise us to learn that the balance of power is very delicately adjusted and, if disturbed by any cause, allows the two partners to revert to a condition of parasite and host.

Best known and longest recorded of such cases is that resulting in the formation of " nodules " on the roots of leguminous and other plants. A bacterium present in most soils all over the world enters the root and multiplies there, giving rise incidentally to gall-like swellings or nodules (p. 99). The bacteria benefit for a time by having access to a supply of starch or other carbohydrates, but the balance of profit is eventually on the side of the plant partner. Much of the nitrogenous or protein material stored up in the bodies of the bacteria as a result of their ability to use the free nitrogen of the air as a source of food is ultimately used by the plant, which thus indirectly taps a source of nitrogenous food otherwise unavailable to green plants.

Of a similar nature, although not so thoroughly known, are the numerous cases of so-called mycorhiza-plants, in which the roots are beset and invaded by fungi whose presence is tolerated or even welcomed by the plant. This association between fungus and root is called *mycorhiza*, and its formation is widespread among Flowering Plants, a fact which need not surprise us if we reflect that the roots of plants are growing in the soil in contact with fungi and other micro-organisms which everywhere abound.

In the root-systems of some plants mycorhiza is frequent; in others it is invariably present. The physiology of the relationship is still imperfectly understood, but, in the case of at least two families of plants, the Orchids and the Heaths, the relation is of an extraordinarily intimate and delicately adjusted kind, and has become so indispensable to the plant that seedlings do not develop properly unless the fungus partner invades the tissues at the appropriate moment.

The story of what is already known of these relationships is one of the most fascinating chapters in recent botanical research, and it is clear that biotic factors of this nature may operate in a very potent way from the point of view of the plant ecologist (p. 114).

Plant Ecology and Plant Geography—It is interesting to trace the connection between *Plant Geography* in the older sense of the term and the more modern branch of botany called *Plant Ecology*.

The plant geographer, like the plant ecologist, is a student of vegetation and they both recognise the association of definite types of vegetation with particular conditions of climate and soil. The object of the plant geographer is primarily to ascertain the nature of the vegetation *floristically*. He wishes to know of what plant species the vegetation of any given area is composed and, by comparison of such records, to study the range of their distribution elsewhere.

The plant ecologist thinks of the plant as a member of a plant society,—the vegetation of which it is a part,—influenced by the world of inanimate nature around it and interacting with other living organisms in the most complex way. He studies the growth and behaviour of the individual plant by the experimental methods of plant physiology. He studies, also, the behaviour of the vegetation unit as a whole and he recognises that a colony or assemblage of plants growing together may undergo changes not unlike those which affect an individual plant, just as the life of a herd of wild animals has a life of its own distinct from that of the animals composing it, or the behaviour of a crowd is very different from that of one of the persons included in it.

Plant Ecology comes into touch with many other branches of Natural Science: with *Geography*, because it studies plants in relation to their physical environment of climate and soil; with *Archæology*, because the settlements of prehistoric man and his movements from one place to another were closely related to the distribution of natural units of vegetation such as marsh, forest and fen, and, when he had reached the pastoral stage, by the position of fertile areas where he could cultivate his crops. The intimate relation of this branch of Botany to the problems of Agriculture and Horticulture has already been noted.

Plant Ecology comes into touch also with Animal Ecology; the predominance of one plant rather than another, for example, may determine the relative abundance, or even the existence, of a member of the local fauna which feeds upon it, or is related to it in other ways. Finally, links are not wanting to bring the former into touch with that very special branch of animal ecology which is called *Sociology*, or the study of human society. Human life, as we know it, depends upon the existence of green plants, and the manner of life of large numbers of human beings is greatly influenced by the fact that their work is concerned with the cultivation of plants.

Practical Work

The following notes as to practical work in plant ecology are designed to suggest other experiments on similar lines. It cannot be too carefully noted that the success of experimental work of this kind depends on the care which is taken in devising methods for carrying out the experiments and upon the accuracy with which all records are kept.

1. Make artificially bared areas of soil in various situations: e.g. in a garden or field; on a gravel path; in turf (a) in the open, (b) beneath trees. Carefully note and record the plants which appear. Areas of this kind should be *at least* one yard square; they can be marked out by pegs at the corners, from which tapes are stretched when records are made. The latter should be kept in the form of permanent records and compared from year to year, drawing any conclusions possible as to the nature of the factors operating.

2. Observe and record the plants and vegetation changes upon accidentally bared areas in different situations: e.g. disused

brickyards where large areas of clay are often exposed; disused gravel pits; molehills and anthills in turf; etc.

3. Make observations upon the plants appearing in 1 and 2, noting especially the means of dispersal of fruits and seeds or the existence of special methods for vegetative reproduction. Make careful notes and drawings of any features of special interest in this connection.

4. Make similar observations to those described in 1 and 2 on areas covered with vegetation and compare at suitable intervals. Permanent pegs should be driven in to mark out the areas, which can then be outlined by tapes; cross-tapes stretched across the enclosed area or *quadrat* render observations more easy.

5. Make observations on the effects produced by grazing animals in a field or on a heath. It is often instructive to protect small areas with wire netting and note the effects upon the plants within them.

6. Fill boxes or pans with soils which bear a special type of vegetation naturally: e.g. peaty soil from a heath, calcareous soil from a chalk down, soil from a salt marsh. Collect seeds of plants which occur naturally on each of the different soils and sow mixtures of them, keeping the seed pans under as natural conditions as possible. Keep records of (*a*) the number of seeds sown, (*b*) the number of seedlings which germinate, (*c*) the subsequent fate of these seedlings.

7. An attempt should be made to map and record the plants on a small area of natural vegetation. Maps for field work can be prepared by making tracings from the six-inch survey map. Records made on these tracings should be carefully compared and transferred finally to the survey map. All vegetation work of this kind should be preceded by careful study of the geology of the area, detailed observations of the nature of the soil, slope, aspect, etc.

CHAPTER XV

THE SOIL

THE NATURE OF SOIL FACTORS. THE CHEMICAL ASPECT OF SOILS:
MINERAL CONSTITUENTS; ORGANIC MATERIAL. CLASSES OF SOILS.
PLANT FOOD IN THE SOIL. THE "CALCIFUGE" AND "CALCICOLE"
HABITS. THE PHYSICAL ASPECT OF SOIL: PORE SPACE; PERCOLATION
AND LIFT OF WATER; SURFACE TENSION AND GRAVITY; CAPILLARITY.
THE SOIL WATER: "TOTAL" AND "AVAILABLE" WATER. THE PROPER-
TIES OF CLAY. COLLOIDS. HUMUS AND CHALK. SUMMARY. THE
BIOLOGICAL ASPECT OF SOIL: BIOLOGICAL SOIL FACTORS; PARTIAL
STERILISATION OF SOIL

The Nature of Soil Factors—The relatively great effects
produced on vegetation by soil factors as compared with those
related to climate in a comparatively small country like
Great Britain have already been noted. Climatic factors such
as temperature, rainfall, humidity and aspect must always
be taken into account in a study of vegetation, but their
effect is often greatly modified by soil differences. Some
knowledge of soil conditions is indispensable to the student,
and we must now study the nature of soil factors and the
ways in which they may influence plants.

Differences in soil conditions may affect the growing plant
directly, but it must not be forgotten that they also act
indirectly by their effect on the growth of myriads of animals
and plants which have their home in the soil. This is especially
true of the many species of bacteria which live and grow in
the soil and, since these may be acted upon directly by other
minute organisms, changes in what is called the *microfauna*
and *microflora* of the soil may in turn affect the vegetation
growing upon it. These direct and indirect effects are in-
timately related to one another and must always be considered together. Bearing this in mind, let us consider some
of the ways in which the soil is related to a green plant whose
roots are growing in it.

It is obvious, in the first place, that such a plant is living in
two worlds,—the shoot system in the air, the root system in the

soil,—and that it can draw its food-supply from either or both of these sources.

We have already learnt in Chapter IV. that a large proportion of the raw materials required as food by green plants is obtained from the soil and is absorbed by the roots in a soluble form. We have learnt, also, that plants require a large and constant supply of water, since mineral substances are absorbed only in very dilute solution, and a large amount of water is required to make good the loss during transpiration and so maintain the cells in a turgid condition.

In the first place, therefore, plants are related to the soil through the *water-supply*, for which reason the water-holding and water-yielding properties of the soil are evidently of the first importance.

Secondly, soil conditions affect the plant through the *air-supply*. We know that all parts of the plant require oxygen in order to breathe. To maintain healthy growth, therefore, the roots must grow in soil which is properly aerated, and this, we shall learn, is closely bound up with the relation of the soil to water.

Thirdly, the *temperature* of the soil is important, and this again is largely determined by its wetness or dryness; so that we find that a *wet* soil is also a *cold* soil and a *dry* soil a *warm* soil, differences which greatly influence the " earliness " or " lateness " of the vegetation in any given district.

Fourthly, the soil is the source of all the food materials required by plants with the exception of carbon dioxide, and although the amount of mineral material required by each plant is extremely small, certain substances are indispensable and must be present in an *available form*, i.e. in a condition such that they can be absorbed by the roots.

From this brief review it may be concluded that we can best learn something of the nature of the soil factors which affect plant growth by studying the soil from three points of view: the *chemical* aspect, in order that we may learn something of the chemical nature of the materials of which soil is composed; the *physical* aspect, since the water-supply of the plant is very closely bound up with the physical character or *texture* of the soil; and the *biological* aspect, which leads to a study of the behaviour of the numerous organisms other than green plants which live and grow in the soil.

THE CHEMICAL ASPECT OF SOIL

Soil is really " rotted rock " and the bulk of it consists of mineral fragments and particles formed by the disintegration of rocks.   Rocks are composed of a large number of different minerals, cemented or consolidated to form a solid mass. Under the action of the weather,—alternations of temperature, frost and rain,—some of these minerals become altered chemically and dissolve in water, so that the solid rock gradually crumbles or disintegrates.   The products of this " weathering " action either collect on the surface of the rock, or are carried away by wind or water and eventually deposited in some other place.   In either case, they accumulate to form a soil.   This process of disintegration can be observed on the surface of the hardest rocks, and the material which collects, whether on the spot or at a distance, speedily acquires a population of animals and plants.   Simple plants like the Lichens find a foothold, and as time goes on and material accumulates, a covering of vegetation is formed. These plants and animals not only *live* in the soil, but also *die* in it, and so add to its bulk the products of their gradual decay.

The framework of the soil, therefore, consists of a mass of inorganic mineral particles of various sizes, mixed with which is a variable amount of organic material derived from the decay of animals and plants.   This organic material, or *humus* as it is called, varies in amount in different soils in relation to the abundance of plant and animal life, and the rate of decay of the organic materials which composed their bodies.   It is brown in colour and is responsible for the dark brown colour of most soils.   Soils which are deficient in humus are usually light in colour.

The Mineral Material of Soil—Rocks are composed of a large number of different minerals, many of which can be found in the soil.   For practical purposes, however, the solid bulk of the soil may be considered as consisting chiefly of three kinds of mineral matter:—*silica, silicates of alumina* and *chalk.*

Silica or *quartz* is a common and abundant rock-forming mineral.   It is hard and resists weathering, and for this reason is usually present in the form of comparatively large angular fragments. Sand is composed almost entirely of such grains of quartz, which form also a high proportion of the solid mineral matter of sandy soils.

The rock-forming minerals which give rise to silicates of alumina are called *felspars* and are widely distributed in rocks. These felspars are compounds of alumina (aluminium oxide) combined with soda, potash or lime. They weather easily, so setting free the quartz and harder minerals embedded among them, undergo certain chemical changes and ultimately, as silicates of alumina combined with soda, potash, lime, or iron, form the greater portion of the finer mineral material of soil, and are the chief mineral constituents of clay.

Chalk is derived from the weathering of calcareous rocks such as limestone and may form a high proportion,—up to over 40 per cent.,—of the soils derived from such rocks. Most of the calcium carbonate or chalk of which calcareous rocks are composed is organic in origin, since these rocks are built up of the remains of vast numbers of animals with calcareous shells or skeletons.

The Organic Material of Soil—By removing the superficial covering of dead leaves and examining the material which collects on the ground beneath the trees in a Beech wood, a good idea can be gained of what is meant by humus. Year by year the leaves fall and slowly decay, giving rise to a dark brown mould filled with remains of twigs, roots and other débris. This mould is almost pure humus and consists of a mixture of organic materials derived from the leaves of successive years in various stages of decay, together with the remains of other organisms which lived in it. The organic substances are compounds of complex chemical nature containing carbon, hydrogen, oxygen and nitrogen, and they represent various stages in the decay of the substances which once formed the tissues of plants and animals. Fresh supplies of organic material are constantly added, the processes of decay are continually going on, and the term " humus " is used to describe the variable and constantly varying mixture thus formed.

Classes of Soils We can make a rough classification of soils based on the proportion of these four constituents. Thus, in *sands and sandy soils*, the proportion of silica as compared with that of silicate of alumina is high; in *clays and clayey soils*, the proportion of the latter is high compared with that of silica. Soils of intermediate type are called *loams* and become sandy loams or clayey loams as the proportion of silica or silicates of alumina respectively increases. If the amount of chalk present in any soil is large (over 10 per cent.) it is described as a *calcareous soil ;* such soils usually have special characters and bear a distinct type of vegetation (p. 212). A clayey soil with much chalk is called a *marl.*

Similarly, if the proportion of humus is high, we get a special class of soils known as *organic soils*, which may be of the nature of *fen* or of *peat*, according to whether calcareous matter is present in any quantity in the soil water or not. As a matter of fact, for reasons which will appear later, calcareous soils are usually deficient in humus while chalk is absent from peaty soils.

Plant Food in the Soil—We have already learnt that terrestrial green plants obtain from the soil, in addition to water, oxygen and nitrates, a number of mineral substances of which compounds of phosphorus, sulphur, potassium, calcium, magnesium and iron are the chief, and that these elements are essential for plant growth. We know that compounds of potassium, calcium, magnesium and iron are formed during the weathering of felspars, and that compounds of phosphorus and sulphur are continually restored to the soil by the decay of organic material,—the unpleasant smell of rotting cabbage or other vegetables, for instance, is due to the formation of a gas named sulphuretted hydrogen, which is a compound of hydrogen and sulphur. We know, also, that the various mineral constituents react with one another and that they are present in the soil as salts,—phosphates, sulphates, nitrates, chlorides, or carbonates,—many of which are soluble and are in a suitable form for absorption by the roots of green plants.

As regards the raw materials of plant food, therefore, most soils contain adequate supplies, and these are added to by the slow weathering of the mineral soil particles. In the case of natural vegetation, the salts removed from the soil by plants during growth are returned to it by the decay of their tissues after death; in the case of cultivated land, they are removed with the crop, a procedure which may result in a deficiency of certain mineral constituents in the soil, more especially of potash and phosphates.* The application of manures by the farmer and gardener is designed to remedy this deficiency, and has two main objects: (1) to make good the loss of mineral matter removed by previous crops, (2) to alter and improve the physical condition of the soil.

When a " natural manure " like farmyard manure is applied to soil, or when animals are allowed to graze upon a field, the cultivator is really restoring some of the mineral matter after its passage through the body of a domestic animal, together with a certain amount of organic material which may have a very considerable effect on the physical character or *texture* of the soil. When " artificial manures " such as " superphosphate," or sulphate of ammonia are used, they serve to make good some special deficiency in mineral constitution, but do not materially affect the physical conditions of the soil *directly*, albeit they may do so *indirectly* if their application is continued over a long period.

The actual chemical composition of the soil is not usually of great significance.† Two exceptions to this general statement must be noted; the first concerned with the amount of calcium carbonate or chalk present, and the second with the presence of large amounts of common salt (sodium chloride) in soil constantly wetted with salt water.

Chalk is a substance of great importance in the soil, affecting the growth of plants both directly and indirectly. Its effect on the soil is discussed elsewhere, but in addition to this it seems to have a direct effect on vegetation, the nature of which is not clearly understood. The vegetation of calcareous soils, i.e. those containing a large percentage of calcium carbonate, is characterised by the absence of certain plant species and the almost invariable presence of others. Of the former may be mentioned Foxglove (*Digitalis purpurea*),

---

* The loss of nitrates is considered in Chapter VI.
† There are a few exceptional cases of soils in very dry climates,—so-called *alkali soils*,—which are unable to support vegetation owing to the amount of alkaline material present. A few cases are known also in which some mineral substance present in the soil appears to be directly associated with peculiarities of the plants growing on it.
A harmful accumulation of salts in the soil may arise where badly drained land is artificially irrigated in districts of low rainfall.

Gorse (*Ulex europæus*), Ling (*Calluna vulgaris*) and Heathers (*Erica spp.*). Whether such plants dislike chalk as such, or do not find otherwise congenial conditions on calcareous soils, is not at present clear; nor is it known whether this *lime-shy* or *calcifuge* habit, as it is called, is of similar nature in plants of different families. As examples of the *lime-loving* or *calcicole* habit may be mentioned Traveller's Joy (*Clematis Vitalba*), the Pasque Flower (*Anemone Pulsatilla*), Lady's Fingers (*Anthyllis Vulneraria*) and Horse-shoe Vetch (*Hippocrepis comosa*); but again there is the same difficulty of knowing whether these plants actually like lime, or whether they are favoured by certain conditions associated with a high proportion of lime in the soil. Many other plant species growing on calcareous soils are found indifferently on dry soils whether with or without a high percentage of chalk. The characteristic features of calcareous soils on the one hand, and of moor or heath soils which are notably deficient in chalk on the other, is the presence in the vegetation of a number of lime-loving species in the first case, and of a majority of lime-hating or lime-shy plants in the second case.

On land periodically overflowed by salt water, the vegetation consists of a relatively small number of species which can endure a high percentage of salt in the soil. Plants growing in such situations usually show structural modifications resembling those found in plants growing in dry situations (p. 55).

THE PHYSICAL ASPECT OF SOIL

The " Texture " of Soil—The " Mechanical Analysis " of Soil—The physical properties of soil depend on its structure and determine what is known to the practical man as its " texture." This is of special importance in connection with the water-supply, and largely determines the amount of water available for a plant or for a crop. It has been found experimentally that many of the properties shown by sand and clay, for example, depend rather upon the *size* of the soil particles than upon their *chemical nature*.

By passing a sample of soil through a series of sieves with meshes of various known sizes, it can be graded into parts or fractions, each of which contains only particles below a certain size. A similar result can be attained for the finer fractions by mixing up a sample of soil with water and allowing the particles to settle, pouring off the liquid into a fresh vessel at stated intervals. Soil chemists have adopted a standard number of fractions into which they separate a sample of soil when making what is called a *mechanical analysis*. They use the methods just described, and the following table shows the way in which the soil materials are classified by soil chemists in this country.

TABLE SHOWING FRACTIONS ADOPTED IN MECHANICAL
ANALYSIS OF SOILS

		Diameter in Millimetres		
		Maximum	*Minimum*	
1	Stones and gravel ..	—	3	Separated by sifting
2	Fine gravel	3	1	
3	Coarse sand	1	0·2	
4	Fine sand	0·2	0·04	Separated by subsidence
5	Silt	0·04	0·01	
6	Fine silt	0·01	0·002	
7	Clay	0·002	—	

Broadly speaking, the proportion of the coarser fractions present determines the physical character of the soil; thus, a sand contains a large proportion of these, a sandy loam a smaller proportion, a clayey loam or a clay still less. Contrariwise, in a clay soil the proportion of finer fractions is great, a majority of the particles in a fine clay having a diameter of ·002 millimetre or less.

Excluding for a moment the effects of a large amount of chalk or humus, and taking into account only the proportion of coarser or finer particles present in a soil, let us consider briefly how the structure of the soil affects the water-supply of the plant.

Pore-Space—A mass of soil consists partly of solid particles packed together more or less closely, partly of spaces between them. The latter collectively form a continuous system of spaces or channels which is known as the *pore-space* of the soil.

If we fill a vessel with shot or marbles, shaking them down so that they are packed as closely as possible, and calculate the volume of the pore-space by measuring the amount of water required to fill the vessel, it will be found that the volumes are the same whether the shot or marbles are large or small, *provided* that all are of the same size in one vessel. The diameter of the channels is less when shot of small size is used, but this is compensated by the greater number of channels present.

In a fine-grained soil like a clay, in which the pore-spaces are much less in diameter than in a sandy soil, *the total pore-space is nevertheless many times greater than in a sandy soil.* This apparently contradictory result is due to the fact that a majority of particles in a clay soil are so minute and so light in weight that they do not arrange themselves so compactly as do the larger and heavier particles of a sand or as does the shot in the experiment just described, and it is known that the volume of the pore-space may be affected by the *arrangement* of the particles although not by their *size*.*

In a clay soil, the total surface of the particles is immensely greater than is that of the larger but fewer particles which make up a like volume of coarse-grained soil or sand. It has been calculated that, roughly, the pore-space of a fine clay amounts to 52·9 per cent. of the total volume, and of a sandy soil to about 32·5 per cent. Correspondingly, the total surface area of the particles in a cubic foot of soil is 173,700 sq. ft. in a clay, but only 11,000 sq. ft. in a sand. It has been estimated that the total surface of all the particles in a cubic foot of an ordinary loam or garden soil would cover an acre if spread out. These values help us to realise the immense area over which the root hairs of a plant, whose roots occupy only a comparatively small region of soil, can be in close contact with the surface of the soil particles. If similar volumes of two soils, one fine-grained, the other coarse-grained, are dried and weighed, the former should weigh less since a smaller proportion of the volume is occupied

* This is strictly true only when all the particles are spheres and are of uniform size in any one case. In soil, the conditions are much more complex since the particles composing it are of different shapes and sizes. Small particles can occupy the spaces between large particles and thus reduce the amount of pore space.

by the pore-space : and this is found to be actually the case. Hence, when we speak of a " heavy clay soil " we must refer to some physical quality other than its actual weight (p. 255).

The Soil Water—Of the water which falls on the soil, some evaporates at once, some runs off the surface, and only a proportion percolates through the soil. After much rain this may occupy the whole of the pore-space so that the soil is *saturated*, and the soil atmosphere which fills normally the pore-spaces completely displaced by water. In a well-drained soil much of the water drains away, leaving a residue, amounting to from 10 per cent. to 20 per cent. of the total weight of water which falls upon the soil, as a film on the outside of the soil particles. This film of water is retained owing to the action of the physical force known as *surface tension*, and its thickness depends upon the interaction of surface tension, which tends to retain the water on the surface of the particles, with the force of gravity which tends to remove it.

The depth to which water percolates is determined by the nature of the underlying rocks and the contours of the ground ; but at some distance below the surface, it may be a few feet or it may be some hundreds of feet, it collects and fills the whole of the pore-space, so forming what is known as the *water-table*, below which the soil or the rock is completely saturated with water. The depth of this water-table below the surface is important to plants since it represents a reservoir upon which they may be able to draw in time of drought.

Percolation and Lift of Water in Soil—Owing to the smaller size of the soil spaces and the greater friction thereby produced, water percolates much more slowly through a fine-grained than through a coarse-grained soil, for which reason clay is liable to become water-logged in wet weather, owing to the difficulty with which water drains away. The reverse is true of a sandy soil, through the comparatively large spaces of which water drains away very rapidly.

It has been stated that under ordinary conditions the soil water is present as films or shells of water on the outside of the particles. These water films form a lining to the pore-spaces everywhere and are in contact throughout; so that we must think of the soil as permeated in all directions by myriads

16

of channels lined with a film of water and extending from the surface of the ground downwards to the water-table.

Surface Tension and Gravity—If there is free drainage in a soil, the films of liquid retained by surface tension on the surface of the particles represent a condition of equilibrium, in which the downward " pull " exerted by gravity is counter-balanced by the force of surface tension.   Hence, the thickness of the films is controlled by the relations between these two forces, since the amount of " pull " exercised by gravity de-pends on the weight or mass of water in the film, and the

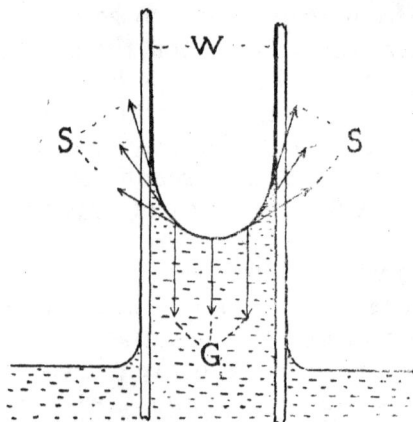

FIG. 34—DIAGRAM TO ILLUSTRATE THE FORCES IN OPERATION WHEN A TUBE IS DIPPED INTO A LIQUID WHICH WETS THE SURFACE

w, film of liquid on inside of tube; s, the arrows indicate the direction in which the force of surface tension is acting, thus tending to raise the liquid in the tube; g, the arrows indicate the downwardly directed " pull " of gravity

" holding force " of surface tension depends upon the surface area.   The matter will perhaps be clearer if we diverge for a moment to consider some of the more marked effects of surface tension.

We may put the matter briefly thus:—the free surface of a liquid like water behaves as though it were a stretched skin, the force of surface tension tending to make it contract. Suppose we take a narrow glass tube and dip the lower end into water.   The water will at once rise some distance up the tube above the level of that outside.   This is because the water, evaporating and condensing on the inside of the tube, wets it; the surface layer, as a result of its tendency to contract, pulls the water up the tube until the weight of the water so raised just counterbalances the "pull" of the surface layer (see fig. 34).

This exhibition of surface tension in narrow tubes, resulting in this case in water rising in the tube, is called *capillarity* (*capillus*, a hair). If a tube having a bore of twice the size is used, the force tending to raise the water will be twice as great owing to the larger surface of water, so that double the volume of water will be raised. This does not result in the water being raised twice as high, however, but to only half the original height, since twice the volume of water occupies only half the length of a tube of double the size. In other words, the smaller the diameter of the tube, the higher the water will rise in it, because although the free *surface* of the water is smaller and therefore the *volume* that can be pulled up correspondingly smaller, the latter occupies a very much *greater length* in the tube. We may put the matter in the converse way and say that the higher the water has to be lifted above the general water surface, the narrower must be the tube.

Let us return now to soil. The soil spaces may be regarded as a continuous network of tubes; the fact that they are not straight tubes does not affect the matter. As we get further and further above the water-table, so we find that only the smaller spaces are filled with water; and, in the upper part of a well-drained soil, the water is present only as a film covering the soil particles. In other words, the amount of water in the soil *decreases* from the water-table upwards.

At or near the surface, another force comes into play, viz., that of *evaporation*, and the water films in the superficial layers of soil tend to become thinner owing to loss from this cause, and also to the constant removal of water by the roots of plants. In this way, equilibrium between the two forces of gravity and surface tension is disturbed, and, in order to restore it and replenish the water films, water rises by capillarity from the layers below which, in their turn, are supplied in a similar way.

As was pointed out on a previous page, water rises higher by capillarity in a tube of narrow diameter than in a wider one; correspondingly, the lifting power, capillary power, or capillarity of a soil depends on the size of the pores, and is greater as the diameter of these is less. Since the average size of the channels is much less in a fine-grained than in a coarse-grained soil, the capillary or lifting power of a clay soil is much greater than that of a sand—i.e. water can be lifted by capillarity from a greater depth. On the other hand, just as water percolates with difficulty through clay soil because of the small size of the water channels, and the immensely greater friction it has to overcome in consequence, so water is lifted by capillarity *more slowly* in such a soil for the same reason, although the final height reached will be greater.

The effectiveness of this lift of water to plants growing in the surface soil depends partly on the depth of the water-table below the surface, and partly upon the physical character of the soil. Thus, in a sandy soil, water is lost rapidly in hot weather, owing to evaporation from the surface layers and the demands of the plant roots. Ability to withstand a long drought on such a soil will depend on the depth of the water-table below the surface. The capillary power is low, water can be lifted but a short distance and, if the water removed from above is not renewed from below, the soil quickly "dries out " and the plants growing upon it soon suffer from drought in a dry season. On the other hand, although a fine-grained soil can lift water from a great depth, the *rate* at which the water rises may be so slow that plants suffer from lack of water, and for this reason one often observes the surface of clay soil cracked and baked in hot weather, although an abundant supply of water exists but a few feet below the surface.

Considered from the point of view of water-supply, therefore, the ideal soil for plants is one of intermediate character, which can lift water from a considerable depth and allow this lift to take place fairly rapidly.

" Available " Water in Soil—Only a certain proportion of the film of water retained on the soil particles is " available " for the plant, and this varies with the nature of the soil. Some soils " hold " water more strongly than others; this means that, although a good deal of water may be retained by a soil, it does not always follow that a large proportion of it can be absorbed by the roots of plants.

The average *total water content* of a soil at any given time can be estimated by taking a known weight of soil, drying it at 100° C. or less, and weighing at intervals until it ceases to lose weight. The difference between the first and last weights will be the weight of water in that sample, from which the total water content of any quantity of soil can be easily calculated. It is not so easy to find out what proportion of this water is *available* for plants, but this can be done roughly by growing a plant in a pot of the same soil and estimating in a similar way the amount of water present when the leaves of the plant flag. The difference between these two values will represent the amount of water available for the plant at the time of the experiment. For reasons discussed in a preceding chapter (p. 50) this method of discovering the amount of *available water* in a soil is not by any means a satisfactory one, and the problem of measuring the proportion of the soil water which can be actually used by a plant at any given moment is very complex and difficult.

The Properties of Clay—Clay possesses remarkable physical properties, due in part to the small size of the constituent particles, and in part to their chemical nature. It is *plastic* when wet and retains its shape when moulded. It holds much water and shrinks greatly when dried,—so much so that it cracks. If a little clay is mixed with water and poured into a tall glass vessel, the larger particles quickly settle down, but the water remains turbid for an indefinite time owing to the presence of fine material which remains in suspension. When dry, clay can be broken up, and after heating or burning does not regain its plastic properties when again wetted, i.e. the materials of which it is composed have changed. It is these properties of clay which affect the " texture " of a clay soil and the ease or difficulty with which it can be worked. Such a soil is tenacious and sticky when wet, holds water strongly, and contracts, so causing cracks on the surface as it dries in hot weather.

These properties depend chiefly on the fact that a high proportion of the finer particles in clay are in the *colloidal* condition. Without attempting to explain here the exact meaning of this statement, the matter is so important that we are justified in trying to form some sort of mental picture of the conditions existing in colloids such as those of clay.

Colloids—When a soluble substance like sugar is added to water it is said to *dissolve*,—by which is meant that it disintegrates into particles at least as small as molecules, which intermix with the molecules of water. When an *insoluble* powder is added to water the particles retain their individuality; if large, they quickly fall to the bottom by their own weight, if small, they fall less quickly. The rate of fall of a particle through water is determined by its weight tending to make it fall, and the friction it encounters in travelling through the water which delays its fall. If the particles are very minute, the rate of fall becomes inappreciable, and they remain suspended in the water. As a matter of fact, when the particles are of such minute size, electrical forces come into play which help to keep them from settling. Such a suspension of very fine particles, so small that they may be invisible through a microscope, exhibiting in some degree the properties of a solution (although behaving very differently in many respects from an ordinary solution), is called a *colloidal solution*.

Colloidal solutions may have the appearance of ordinary liquids except that they appear opalescent or turbid; often they form a jelly or " gel " upon the addition of small quantities of certain salts. It is probable that in soils containing much colloid material the film surrounding the soil particles may be in the condition of jelly rather than that of liquid. Substances in the colloidal state possess other remarkable properties and may exert profound effects upon solutions with which they are in contact. Among those of importance in connection with soil colloids are the power of withdrawing substances from solutions and retaining them (adsorption), and the power of taking up and holding large quantities of water. The colloidal material in soil is derived both from the weathering of mineral material and from humus. Many of the characteristic physical properties of clay and peat soils are due to the high proportion of colloids they contain, derived from these two sources respectively.

Flocculation—Although the fine particles of a colloidal solution, such as clay, will not of themselves settle, it is possible to make them do so by the addition of certain chemical substances. The addition of lime-water, for example, to a clay suspension causes a rapid clearing of the turbid solution and settling of the particles. This effect is brought about by conversion of the lime-water to soluble bicarbonate of lime, which then causes the colloid particles to aggregate and thus form groups heavy enough to settle.

Substances which produce such an effect are said to *flocculate* the clay suspension: alkalis in general produce a reverse effect and tend to keep the particles in suspension. A similar effect is produced by the addition of lime or chalk to a clay soil; this, by its flocculating effect, may profoundly modify the properties of such a soil.

Humus and Chalk—The presence of considerable amounts of humus or chalk may profoundly modify the physical properties of the soils. The reasons for this can only be very briefly indicated here, and since they are to some extent related, it is convenient to consider them together.

The processes of decay responsible for the formation of humus are brought about by micro-organisms,—chiefly Bacteria,—and the general character of the changes which take place has already been studied (p. 96). Most of the Bacteria which carry on the later stages of decomposition of organic matter are aerobes and require the soil to be well aerated. Any conditions which hinder free aeration, as, for instance, bad drainage with the water-logged condition of the soil resulting from it, hinder their activities or bring them to an end. The earlier products of decay accumulate; many of them are of an acid nature and tend to make the soil " sour " and unsuitable for the growth of most plants. If the process continues, peat is formed, and since only certain plants will grow on an acid peat soil, it soon acquires a special type of vegetation, or, if under cultivation, ceases to grow good crops.

Humus contains great stores of nitrogen and carbon " locked up " in the form of organic substances of complex chemical constitution. These nitrogenous materials cannot ordinarily be used as food by the higher plants until they have undergone further changes due to the action of Bacteria and reached the condition of nitrates (p. 96). The addition of humus to soils in which it is deficient is desirable because of its physical effect upon the soil, and because it supplies the raw material from which nitrates are formed. The processes responsible for the conversion of humus to the form of nitrates are carried on in successive stages by soil bacteria, which use the organic materials as food, and by so doing break them down into simpler substances, which in turn are used by the nitrifying bacteria and so rendered available to the higher plants. This wonderful cycle of changes is continually going on in soil humus if conditions are favourable to the bacteria concerned. Of such conditions, one of the most important is good aeration of the soil, without which the processes of decay speedily come to an end.

It so happens that the " acid " properties of these organic products of decay are neutralised by chalk or lime, which can thus " sweeten " a " sour " soil and so facilitate fresh bacterial activity. These facts enable us to understand why peaty soils are deficient in lime and why calcareous soils contain but a small proportion of humus. In the first case, the processes of decay are brought to an end by the accumulation of products which hinder or prevent bacterial activity; in the second case, such substances are immediately neutralised, bacterial activity is encouraged, and any humus present rapidly undergoes decay.

Chalk has an important physical effect on clay soils. Owing to its "flocculating" effect on the colloidal material of such soils, it tends to make them more friable in texture and therefore more easy to cultivate. It must not be forgotten that chalk or lime supplies an essential constituent of plant food, and that it is liable to be washed out of soil rather rapidly by rain, whereby the soil may become deficient in calcareous material.

The nitrogen and carbon "locked up" in humus are not in such forms that they can be utilised by green plants as food, and for this reason humus is chiefly important because of its physical effect upon the soil, and its indirect effect as a source of food for the soil fauna and flora. *If sufficient chalk is present*, the addition of humus to clay soils tends to improve their texture and makes them more easy to work; when added to sandy soils, it increases their water-holding properties, owing to the fact that much of the organic material is in the colloidal condition.

Summary of Physical and Chemical Aspects—Let us summarise what has been learnt concerning the physical and chemical aspects of the soil.

Soil consists of a framework of relatively large mineral particles derived from the decomposition of rocks, associated with which is a variable proportion of particles or groups of particles of colloidal nature, partly of mineral origin derived chiefly from the weathering of silicates of alumina, and partly of organic origin derived from humus. The proportion of such colloidal material varies greatly in different soils and may profoundly affect the soil in relation to plant growth. As regards the raw mineral material of plant food, most soils contain sufficient supplies, and the weathering of the mineral soil particles goes on continuously although slowly. The water-supply of plants is determined chiefly by the physical structure of the soil. Hence, for practical purposes, it is often more valuable to the cultivator of plants to understand how he can modify this by cultivation than to possess a knowledge of the chemical composition of the soil.

Enough has been said to show how very complex are the problems of soil physics, and to indicate broadly how the physical structure of the soil is related to the water-supply of plants. The problems involved are extremely complex and the space at our disposal permits us to consider them only in outline. Treated even in this cursory manner, it is clear that most of the cultural operations performed by farmers and gardeners have for their object the aeration of the soil and the control of the water-supply.

Thus, *ploughing* and *digging* loosen the soil, improve the drainage and give free access to air; *hoeing* the surface breaks the continuity of the capillary channels and so checks loss of water by evaporation, since the water reserve below is not drawn upon to take the place of that lost. A *mulch* of straw or other material serves the same purpose by *preventing* evaporation from the surface; *rolling*, on the contrary, compresses the surface soil, thus increasing its capillarity and facilitating loss of water by evaporation; it is therefore only carried out as a temporary measure for increasing the supply of water to the roots of young plants growing in the superficial layers of soil in the spring or early summer. Mention has already been made of the striking effect produced on natural vegetation by consolidation of the soil and the improvement in capillary properties resulting therefrom.

THE BIOLOGICAL ASPECT OF SOIL

The study of soil from the biological point of view is of so vast extent that to record what is known in detail would require a whole book. It is only possible here to give some idea of its great importance.

The soil is the home of a multitude of living things, all carrying on separate life-processes of nutrition, growth and reproduction. Many of these life-processes interact with one another and with those of the higher plants in the most complex and wonderful manner.

The ways in which members of the soil population can behave directly as *biotic factors* in the environment of the plant have already been explained, and we will now briefly review the facts relating to effects produced indirectly by minute organisms living in the soil. The part played by soil Bacteria and Fungi in the processes of decay, whereby complex organic materials, composing the bodies of plants and animals, are resolved after death into simple constituents such as water, carbon dioxide and ammonia, has already been learnt. Linking these processes with the growth of the higher plants are the activities of the *nitrifying bacteria* and of the *nitrogen-fixing bacteria*. These organisms form indispensable links in the chain of events by which Flowering Plants, and through them the higher animals and man, are provided with nitrogen in forms such as they can readily use. The green plant thus obtains supplies of nitrates and, by uniting them with carbon compounds, forms proteins which help to build up new protoplasm, or protein reserves. Much of this plant protein is eaten by animals and can ultimately serve, either directly in the form of *vegetables* or *cereals*, or indirectly in the form of *meat*, as food for human beings.

It is difficult to make direct observations on the behaviour of soil bacteria, and there is still much to be learnt regarding their manner of life and the complicated chemical changes to which they give rise in the soil. There is the problem, for example, presented by the existence of anaerobic bacteria in the soil. How, it may be asked, can organisms to which oxygen is poisonous find a congenial home in a well aerated soil? Two solutions of this problem may be suggested, both the result of experimental work on soil problems.

One presents itself as a result of work on the soil bacterium *Clostridium*, which, although an anaerobe, can live in soil if closely associated with certain aerobic bacteria which use up the oxygen near it (p. 99). Another explanation can be deduced from a recent investigation into the conditions existing in the films of liquid which surround the soil particles. Various gases are dissolved in this liquid as they would be in water or a watery solution exposed to the air elsewhere. Of these gases, carbon dioxide and nitrogen are the chief, and oxygen is practically absent. These water films are doubtless the dwelling-place of countless minute organisms which, living below the surface of the liquid, are thus living in a world practically free from oxygen.

We have hitherto assumed the existence in soil of organisms of microscopic size. It is not difficult to satisfy ourselves by a simple experiment that this is actually the case. If a little moist soil is taken and placed under such conditions that we can control and examine the air as it enters and leaves the vessel in which the soil is contained, it can readily be shown that oxygen is removed from the air after entering the vessel and that carbon dioxide is given off from the soil—i.e., that *respiration* changes are taking place in it. If we then heat the same or a similar sample of soil to a temperature fatal to life and repeat the experiment, it can be proved that these changes no longer take place, and therefore must have been due to the respiration of micro-organisms (p. 255, experiment 11).

Other facts which throw light on the complicated nature of the biological activities of soil have also been discovered by experiments on the effect of heating soil, or otherwise treating it so as to destroy all or some of its living inhabitants. If soil is exposed to a high temperature, 120° C. or more, for an hour or less, all life will be destroyed: it will be *completely sterilised*. Soil treated in this way, however, undergoes other chemical changes which render it unsuitable for the growth of plants. If the temperature of boiling water is used, or if soil is exposed to the action of antiseptic substances, e.g. carbon disulphide, which are quickly fatal to some forms of life and, being volatile, can be completely removed after the experiment, some organisms are killed outright but others survive. By means of such experiments it has been shown that soils *partially sterilised* in this way become more fertile for ordinary plants, and also that soils which have been over-manured or in other ways become unsuited for plant growth can be restored to fertility.   One explanation of these results is believed to be that, under certain conditions, some kinds of soil micro-organisms increase rapidly at the expense of others, and so prevent or hinder processes which are of benefit indirectly to the Higher Plants.   By the methods of partial sterilisation described, the unfavourable organisms are destroyed, while those carrying on processes favourable to the growth of plants are preserved and encouraged.

By such experiments do we learn that *biotic factors* in the environment may play as important a part in the ecology of the lowly forms of life which inhabit the soil as they do in the ecology of the Higher Plants.

Practical Work

1. Roughly weigh equal quantities of fresh sand, garden soil and clay.   Mix up separately with distilled water, pour into tall glass vessels and stand on a steady table or bench where they will not be disturbed or shaken.   Note the time taken for the particles to settle and the water to become clear in each case.

2. Repeat above experiment, using *two* vessels to which a little clay, rubbed up in distilled water, has been added.   To one vessel add 50 c.c. freshly prepared lime-water and note the " flocculating " effect on the clay particles.

3. Fill two vessels of like volume with shot of two sizes.   (The two sizes of shot known commercially as " dust " and " number 16 " serve well for this experiment.)   Pour water from a graduated glass cylinder into the vessels containing the shot, and measure the volume required in each case to completely fill the pore-space with water.   The result may be compared with that obtained by using an equal volume of sand.

4. The standard method of carrying out a mechanical analysis of soil by means of sieves with meshes of different standard sizes may be learnt from a textbook of soil chemistry.

The apparatus shown in fig. 35 may be used to illustrate the presence of particles of different sizes in soil.   Lead a slow stream of water through the tube A, as shown in the diagram.   Mix up with water in a beaker a small quantity of soil, stir, and add slowly to the water entering the apparatus at B.

The particles present will be deposited in the tube in a graded series with the finest particles nearest to the exit C.

5. Set up as follows an experiment to compare the degree and rate of capillary rise of water in (*a*) coarse sand, (*b*) powdered

garden soil, (c) powdered clay.   Spread out a quantity of fresh
garden soil and of clay on paper in a dry place.   When air-dry,
powder finely.   Prepare a number of pieces of glass tubing of one
inch diameter, cut into lengths of not less than twenty-four inches.
Tie or wire pieces of coarse muslin over one end of each tube.
(A more satisfactory way of fixing the muslin is to cut in a cork a
hole of such a size that the glass tube will fit tightly into it:
the muslin is laid on the cork and the tube pushed into the hole.)

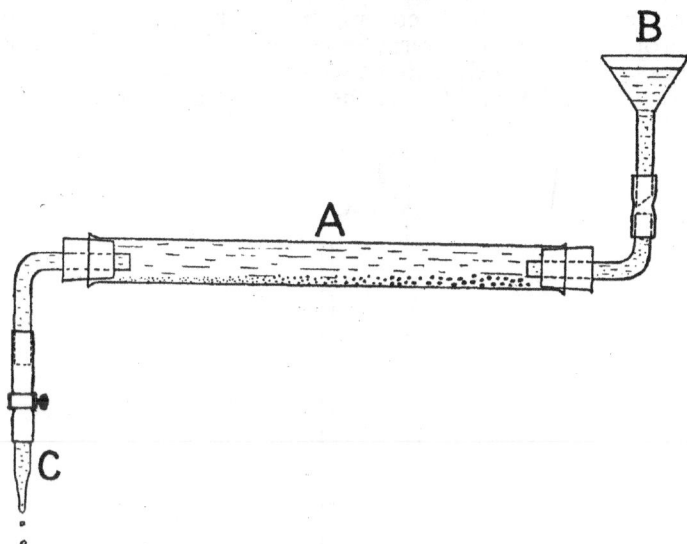

FIG. 35—SIMPLE APPARATUS FOR GRADING SOIL PARTICLES ACCORD-
ING TO SIZE BY THE METHOD OF SEDIMENTATION

The tube A should be at least three feet in length.   (See text)

Fill the tubes with the sand, powdered soil, and powdered clay
respectively and shake down firmly.   Fix the tubes in an upright
position, with the covered ends dipping into vessels of water.
Record the rate at which the water rises in each, and, by using
measured quantities of water, estimate in each case the time taken
to lift a given quantity.

6. It must be noted that the results of the last experiment
give no direct information as to the actual rate of rise of water
in the different soils as they occur in nature.   It is not satisfactory
to use soil in the fresh condition for filling glass tubes, as described
in experiment 5 owing to the difficulty of packing the soils
uniformly.   The results of the experiment can be confirmed and
extended by using columns of soil obtained as follows.   Obtain
cylinders of thin, sheet, galvanised iron, nine inches to twelve inches
in length, fitted with detachable covers at each end.   Remove a
" bore " of garden soil, clay, or peat by forcing a cylinder into a

suitable surface of each. (It is most convenient to prepare a lateral surface by digging a shallow hole.) Clay can often be readily obtained from a brickworks. Although rather troublesome to cut out in the first instance, such a cylinder of clay can be kept moist and used for a large number of experiments to illustrate the behaviour of clay soils to water. Very instructive information as to the water-lifting power, rate of evaporation, and the effect of hoeing or mulching the upper layers of soil can be obtained by the use of these cylinders. When brought into the laboratory, remove both covers, stand the cylinders on small pieces of cork or other material in vessels of water, and make careful measurements of the rate at which water is removed from the vessels, and of the condition of the evaporating surfaces.

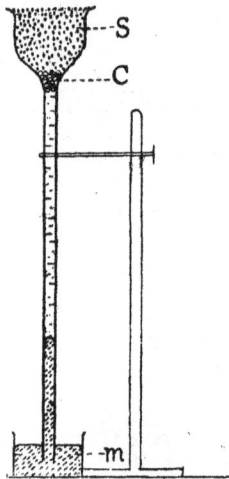

FIG. 36—APPARATUS FOR DEMONSTRATING THE CAPILLARY ACTION OF SOIL ON A COLUMN OF WATER

s, soil; *c*, plug of cotton-wool; *m*, mercury

7. In order to compare the " capillary pull " of various soils, the following experiment is useful. Obtain several thistle funnels with tubes about twelve inches long. Fill the bulbs of four of these carefully and uniformly with fresh sand, garden soil, clay, and peat respectively, pressing gently so as to fill all the large air spaces (fig. 36). (It may be necessary to plug the top of the sand tube with a small piece of cotton-wool.) Smooth off the sand, clay, etc. carefully on the top, invert the funnel and fill the tubes with water, freshly boiled to remove air. The operation of filling the tubes can be performed readily by the aid of a piece of glass tubing or a test tube drawn out to a fine tube in a gas flame. Place the end of each tube in a small vessel of mercury; note and compare the rate of rise and the total height reached by the

mercury in each case. Instead of using mercury, the tubes may be placed with their ends in water, and the capillary powers of the soils compared by measuring the rate at which water is removed from each vessel.

8. Using similar tubes to those in experiment 5, determine the rate of percolation through various soil materials as follows. Prepare several glass tubes as in experiment 5. Place in them equal volumes of the soil materials to be compared, using them dry and powdered as described. Shake down firmly and compare the rate of percolation of 50 c.c. of water through each column, by taking the time of (a) the first drop through, (b) the last drop through. (It may be necessary to fix an arbitrary time-limit for (b), e.g. one drop in three minutes or other time unit.) This experiment can also be performed using shot, as in experiment 3.

9. Repeat the last experiment, measuring and comparing the amount of water retained in each case. This illustrates the different degrees of retentiveness due to differences in size of the component soil particles, i.e. it demonstrates the larger surface over which surface tension is acting in a finer-grained soil.

10. Using small quantities of dry sand, powdered soil and powdered clay, compare the *weights* of equal *volumes* of these soil materials.

11. The following simple experiment serves to demonstrate the presence of living micro-organisms in soil (p. 251). Place a little fresh soil in the bottom of a flask. Fit the flask with a two-hole rubber cork, passing through one hole a piece of glass tubing through which air enters the flask, and through the other a piece of glass tubing bent twice at right angles, leading into a tall, corked vessel containing fresh lime-water. Air enters the flask, oxygen is used up and carbon dioxide is given off during the respiration of the micro-organisms present. By fitting a third glass tube to the vessel containing lime-water and attaching it to an aspirator, the respired air can be drawn through the lime-water, in which the carbon dioxide will cause to be precipitated a fine deposit of calcium carbonate (chalk). That this change in the constitution of the air which has been in contact with the soil is due to the respiration of living organisms is demonstrated by setting up a *control* flask, identical with the other except that the soil has been treated so as to destroy all life. This can be done effectively by taking the flask containing the soil, closed with a plug of cotton-wool, and placing it in a steamer, or standing it in a vessel of boiling water for half an hour on three successive days.

Readers who wish to study the problems of soil further, will find the following books useful:

" Soil Conditions and Plant Growth." E. J. Russell. Longmans, Green and Co.
" The Soil." A. D. Hall. John Murray.

INDEX

Page numbers in italics refer to practical work ; numbers with asterisks refer to text figures.

17   257

A SELECTION FROM
MESSRS. METHUEN'S
PUBLICATIONS

This Catalogue contains only a selection of the more important books published by Messrs. Methuen. A complete catalogue of their publications may be obtained on application.

Bain (F. W.)—
A DIGIT OF THE MOON: A Hindoo Love Story. THE DESCENT OF THE SUN: A Cycle of Birth. A HEIFER OF THE DAWN. IN THE GREAT GOD'S HAIR. A DRAUGHT OF THE BLUE. AN ESSENCE OF THE DUSK. AN INCARNATION OF THE SNOW. A MINE OF FAULTS. THE ASHES OF A GOD. BUBBLES OF THE FOAM. A SYRUP OF THE BEES. THE LIVERY OF EVE. THE SUBSTANCE OF A DREAM. *All Fcap. 8vo. 5s. net.* AN ECHO OF THE SPHERES. *Wide Demy. 12s. 6d. net.*

Balfour (Graham). THE LIFE OF ROBERT LOUIS STEVENSON. *Fifteenth Edition. In one Volume. Cr. 8vo. Buckram, 7s. 6d. net.*

Belloc (H.)—
PARIS, 8s. 6d. net. HILLS AND THE SEA, 6s. net. ON NOTHING AND KINDRED SUBJECTS, 6s. net. ON EVERYTHING, 6s. net. ON SOMETHING, 6s. net. FIRST AND LAST, 6s. net. THIS AND THAT AND THE OTHER, 6s. net. MARIE ANTOINETTE, 18s. net. THE PYRENEES, 10s. 6d. net.

Bloemfontein (Bishop of). ARA CŒLI: AN ESSAY IN MYSTICAL THEOLOGY. *Seventh Edition. Cr. 8vo. 5s. net.*
FAITH AND EXPERIENCE. *Third Edition. Cr. 8vo. 5s. net.*
THE CULT OF THE PASSING MOMENT. *Fourth Edition. Cr. 8vo. 5s. net.*
THE ENGLISH CHURCH AND REUNION. *Cr. 8vo. 5s. net.*
SCALA MUNDI. *Cr. 8vo. 4s. 6d. net.*

Chesterton (G. K.)—
THE BALLAD OF THE WHITE HORSE. ALL THINGS CONSIDERED. TREMENDOUS TRIFLES. ALARMS AND DISCURSIONS. A MISCELLANY OF MEN. *All Fcap. 8vo. 6s. net.* WINE, WATER, AND SONG. *Fcap. 8vo. 1s. 6d. net.*

Clutton-Brock (A.). WHAT IS THE KINGDOM OF HEAVEN? *Fourth Edition. Fcap. 8vo. 5s. net.*
ESSAYS ON ART. *Second Edition. Fcap. 8vo. 5s. net.*

Cole (G. D. H.). SOCIAL THEORY. *Cr. 8vo. 5s. net.*

Conrad (Joseph). THE MIRROR OF THE SEA: Memories and Impressions. *Fourth Edition. Fcap. 8vo. 6s. net.*

Einstein (A.). RELATIVITY: THE SPECIAL AND THE GENERAL THEORY. Translated by ROBERT W. LAWSON. *Cr. 8vo. 5s. net.*

Fyleman (Rose.). FAIRIES AND CHIMNEYS. *Fcap. 8vo. Sixth Edition. 3s. 6d. net.*
THE FAIRY GREEN. *Third Edition. Fcap. 8vo. 3s. 6d. net.*

Gibbins (H. de B.). INDUSTRY IN ENGLAND: HISTORICAL OUTLINES. With Maps and Plans. *Tenth Edition. Demy 8vo. 12s. 6d. net.*
THE INDUSTRIAL HISTORY OF ENGLAND. With 5 Maps and a Plan. *Twenty-seventh Edition. Cr. 8vo. 5s.*

Gibbon (Edward). THE DECLINE AND FALL OF THE ROMAN EMPIRE. Edited, with Notes, Appendices, and Maps, by J. B. BURY. Illustrated. *Seven Volumes. Demy 8vo. Illustrated. Each 12s. 6d. net. Also in Seven Volumes. Cr. 8vo. Each 7s. 6d. net.*

Glover (T. R.). THE CONFLICT OF RELIGIONS IN THE EARLY ROMAN EMPIRE. *Eighth Edition. Demy 8vo. 10s. 6d. net.*
POETS AND PURITANS. *Second Edition. Demy 8vo. 10s. 6d. net.*
FROM PERICLES TO PHILIP. *Third Edition. Demy 8vo. 10s. 6d. net.*
VIRGIL. *Fourth Edition. Demy 8vo. 10s. 6d. net.*
THE CHRISTIAN TRADITION AND ITS VERIFICATION. (The Angus Lecture for 1912.) *Second Edition. Cr. 8vo. 6s. net.*

Grahame (Kenneth). THE WIND IN THE WILLOWS. *Tenth Edition. Cr. 8vo. 7s. 6d. net.*

Hall (H. R.). THE ANCIENT HISTORY OF THE NEAR EAST FROM THE EARLIEST TIMES TO THE BATTLE OF SALAMIS. Illustrated. *Fourth Edition. Demy 8vo. 16s. net.*

Hobson (J. A.). INTERNATIONAL TRADE: AN APPLICATION OF ECONOMIC THEORY. *Cr. 8vo. 5s. net.*
PROBLEMS OF POVERTY: AN INQUIRY INTO THE INDUSTRIAL CONDITION OF THE POOR. *Eighth Edition. Cr. 8vo. 5s. net.*
THE PROBLEM OF THE UNEMPLOYED: AN INQUIRY AND AN ECONOMIC POLICY. *Sixth Edition. Cr. 8vo. 5s. net.*

GOLD, PRICES AND WAGES : With an
Examination of the Quantity Theory.
Second Edition. Cr. 8vo. 5s. net.
TAXATION IN THE NEW STATE.
Cr. 8vo. 6s. net.
Holdsworth (W. S.). A HISTORY OF
ENGLISH LAW. *Vol. I., II., III.,
Each Second Edition. Demy 8vo. Each
15s. net.*
Inge (W. R.). CHRISTIAN MYSTICISM.
(The Bampton Lectures of 1899.) *Fourth
Edition. Cr. 8vo. 7s. 6d. net.*
Jenks (E.). AN OUTLINE OF ENG-
LISH LOCAL GOVERNMENT. *Fourth
Edition.* Revised by R. C. K. Ensor. *Cr.
8vo. 5s. net.*
A SHORT HISTORY OF ENGLISH
LAW : From the Earliest Times to
the End of the Year 1911. *Second
Edition, revised. Demy 8vo. 12s. 6d. net.*
Julian (Lady) of Norwich. REVELA-
TIONS OF DIVINE LOVE. Edited by
Grace Warrack. *Seventh Edition. Cr.
8vo. 5s. net.*
Keats (John). POEMS. Edited, with Intro-
duction and Notes, by E. de Sélincourt.
With a Frontispiece in Photogravure.
Third Edition. Demy 8vo. 10s. 6d. net.
Kipling (Rudyard). BARRACK-ROOM
BALLADS. 205*th Thousand. Cr. 8vo.
Buckram, 7s. 6d. net. Also Fcap. 8vo.
Cloth, 6s. net ; leather, 7s. 6d. net.*
Also a Service Edition. *Two Volumes.
Square fcap. 8vo. Each 3s. net.*
THE SEVEN SEAS. 152*nd Thousand.
Cr. 8vo. Buckram, 7s. 6d. net. Also Fcap.
8vo. Cloth, 6s. net ; leather, 7s. 6d. net.*
Also a Service Edition. *Two Volumes.
Square fcap. 8vo. Each 3s. net.*
THE FIVE NATIONS. 126*th Thousand.
Cr. 8vo. Buckram, 7s. 6d. net. Also Fcap.
8vo. Cloth, 6s. net ; leather, 7s. 6d. net.*
Also a Service Edition. *Two Volumes.
Square fcap. 8vo. Each 3s. net.*
DEPARTMENTAL DITTIES. 94*th Thou-
sand. Cr. 8vo. Buckram, 7s. 6d. net.
Also Fcap. 8vo. Cloth, 6s. net ; leather,
7s. 6d. net.*
Also a Service Edition. *Two Volumes.
Square fcap. 8vo. Each 3s. net.*
THE YEARS BETWEEN. *Cr. 8vo.
Buckram, 7s. 6d. net. Also on thin paper.
Fcap. 8vo. Blue cloth, 6s. net ; Limp
lambskin, 7s. 6d. net.*
Also a Service Edition. *Two Volumes.
Square fcap. 8vo. Each 3s. net.*
HYMN BEFORE ACTION. Illuminated.
Fcap. 4to. 1s. 6d. net.
RECESSIONAL. Illuminated. *Fcap. 4to.
1s. 6d. net.*
TWENTY POEMS FROM RUDYARD
KIPLING. 360*th Thousand. Fcap. 8vo.
1s. net.*
Lamb (Charles and Mary). THE COM-
PLETE WORKS. Edited by E. V. Lucas.
*A New and Revised Edition in Six Volumes.
With Frontispieces. Fcap. 8vo. Each 6s. net.*

The volumes are :—
I. Miscellaneous Prose. II. Elia and
the Last Essay of Elia. III. Books
for Children. IV. Plays and Poems.
V. and VI. Letters.
Lankester (Sir Ray). SCIENCE FROM
AN EASY CHAIR. Illustrated. *Thirteenth
Edition. Cr. 8vo. 7s. 6d. net.*
SCIENCE FROM AN EASY CHAIR.
Illustrated. *Second Series. Third Edition.
Cr. 8vo. 7s. 6d. net.*
DIVERSIONS OF A NATURALIST.
Illustrated. *Third Edition. Cr. 8vo.
7s. 6d. net.*
SECRETS OF EARTH AND SEA. *Cr.
8vo. 8s. 6d. net.*
Lodge (Sir Oliver). MAN AND THE
UNIVERSE : A Study of the Influence
of the Advance in Scientific Know-
ledge upon our Understanding of
Christianity. *Ninth Edition. Crown 8vo.
7s. 6d. net.*
THE SURVIVAL OF MAN : A Study in
Unrecognised Human Faculty. *Seventh
Edition. Cr. 8vo. 7s. 6d. net.*
MODERN PROBLEMS. *Cr. 8vo. 7s. 6d.
net.*
RAYMOND ; or Life and Death. Illus-
trated. *Twelfth Edition. Demy 8vo. 15s.
net.*
THE WAR AND AFTER : Short Chap-
ters on Subjects of Serious Practical
Import for the Average Citizen in A.D.
1915 Onwards. *Eighth Edition. Fcap.
8vo. 2s. net.*
Lucas (E. V.).
THE LIFE OF CHARLES LAMB, 2 *vols.*, 21s.
net. A Wanderer in Holland, 10s. 6d. *net.*
A Wanderer in London, 10s. 6d. *net.*
London Revisited, 10s. 6d. *net.* A Wan-
derer in Paris, 10s. 6d. *net* and 6s. *net.* A
Wanderer in Florence, 10s. 6d. *net.*
A Wanderer in Venice, 10s. 6d. *net.* The
Open Road : A Little Book for Wayfarers,
6s. 6d. *net* and 7s. 6d. *net.* The Friendly
Town : A Little Book for the Urbane, 6s.
net. Fireside and Sunshine, 6s. *net.* The
Character and Comedy, 6s. *net.* The
Gentlest Art : A Choice of Letters by
Entertaining Hands, 6s. 6d. *net.* The
Second Post, 6s. *net.* Her Infinite
Variety : A Feminine Portrait Gallery, 6s.
net. Good Company : A Rally of Men, 6s.
net. One Day and Another, 6s. *net.*
Old Lamps for New, 6s. *net.* Loiterer's
Harvest, 6s. *net.* Cloud and Silver, 6s.
net. Listener's Lure : An Oblique Nar-
ration, 6s. *net.* Over Bemerton's : An
Easy-Going Chronicle, 6s. *net.* Mr. Ingle-
side, 6s. *net.* London Lavender, 6s. *net.*
Landmarks, 6s. *net.* A Boswell of
Baghdad, and other Essays, 6s. *net.*
'Twixt Eagle and Dove, 6s. *net.* The
Phantom Journal, and other Essays and
Diversions, 6s. *net.* The British School :
An Anecdotal Guide to the British Painters
and Paintings in the National Gallery, 6s. *net.*

McDougall (William). AN INTRODUC-
TION TO SOCIAL PSYCHOLOGY.
Fifteenth Edition. Cr. 8vo. 7s. 6d. net.
BODY AND MIND : A HISTORY AND A
DEFENCE OF ANIMISM. *Fourth Edition.*
Demy 8vo. 12s. 6d. net.

Maeterlinck (Maurice)—
THE BLUE BIRD : A Fairy Play in Six Acts,
6s. *net.* MARY MAGDALENE ; A Play in
Three Acts, 5s. *net.* DEATH, 3s. 6d. *net.*
OUR ETERNITY, 6s. *net.* THE UNKNOWN
GUEST, 6s. *net.* POEMS, 5s. *net.* THE
WRACK OF THE STORM, 6s. *net.* THE
MIRACLE OF ST. ANTHONY : A Play in One
Act, 3s. 6d. *net.* THE BURGOMASTER OF
STILEMONDE : A Play in Three Acts, 5s.
net. THE BETROTHAL ; or, The Blue Bird
Chooses, 6s. *net.* MOUNTAIN PATHS, 6s.
net.

Milne (A. A.). THE DAY'S PLAY. THE
HOLIDAY ROUND. ONCE A WEEK. *All*
Cr. 8vo. 7s. net. NOT THAT IT MATTERS.
Fcap. 8vo. 6s. net.

Oxenham (John)—
BEES IN AMBER ; A Little Book of Thought-
ful Verse. ALL'S WELL : A Collection of
War Poems. THE KING'S HIGH WAY. THE
VISION SPLENDID. THE FIERY CROSS.
HIGH ALTARS : The Record of a Visit to
the Battlefields of France and Flanders.
HEARTS COURAGEOUS. ALL CLEAR !
WINDS OF THE DAWN. *All Small Pott*
8vo. Paper, 1s. 3d. net ; cloth boards, 2s.
net. GENTLEMEN—THE KING, 2s. net.

Petrie (W. M. Flinders). A HISTORY
OF EGYPT. Illustrated. *Six Volumes.*
Cr. 8vo. Each 9s. net.
VOL. I. FROM THE 1ST TO THE XVITH
DYNASTY. *Ninth Edition.* 10s. 6d. net.
VOL. II. THE XVIITH AND XVIIITH
DYNASTIES. *Sixth Edition.*
VOL. III. XIXTH TO XXXTH DYNASTIES.
Second Edition.
VOL. IV. EGYPT UNDER THE PTOLEMAIC
DYNASTY. J. P. MAHAFFY. *Second Edition.*
VOL. V. EGYPT UNDER ROMAN RULE. J. G.
MILNE. *Second Edition.*
VOL. VI. EGYPT IN THE MIDDLE AGES.
STANLEY LANE POOLE. *Second Edition.*
SYRIA AND EGYPT, FROM THE TELL
EL AMARNA LETTERS. Cr. 8vo.
5s. net.
EGYPTIAN TALES. Translated from the
Papyri. First Series, IVth to XIITH Dynasty.
Illustrated. *Third Edition.* Cr. 8vo.
5s. net.
EGYPTIAN TALES. Translated from the
Papyri. Second Series, XVIIITH to XIXTH
Dynasty. Illustrated. *Second Edition.*
Cr. 8vo. 5s. net.

Pollard (A. F.). A SHORT HISTORY
OF THE GREAT WAR. With 19 Maps.
Second Edition. Cr. 8vo. 10s. 6d. net.

Price (L. L.). A SHORT HISTORY OF
POLITICAL ECONOMY IN ENGLAND
FROM ADAM SMITH TO ARNOLD
TOYNBEE. *Ninth Edition.* Cr. 8vo.
5s. net.

Reid (G. Archdall). THE LAWS OF
HEREDITY. *Second Edition. Demy 8vo.*
£1 1s. net.

Robertson (C. Grant). SELECT STAT-
UTES, CASES, AND DOCUMENTS,
1660-1832. *Third Edition. Demy 8vo.*
15s. net.

Selous (Edmund). TOMMY SMITH'S
ANIMALS. Illustrated. *Eighteenth Edi-
tion. Fcap. 8vo.* 3s. 6d. net.
TOMMY SMITH'S OTHER ANIMALS.
Illustrated. *Eleventh Edition. Fcap. 8vo.*
3s. 6d. net.
TOMMY SMITH AT THE ZOO. Illus-
trated. *Fourth Edition. Fcap. 8vo.*
2s. 9d.
TOMMY SMITH AGAIN AT THE ZOO.
Illustrated. *Second Edition. Fcap. 8vo.*
2s. 9d.
JACK'S INSECTS. Illustrated. Cr. 8vo. 6s.
net.
JACK'S INSECTS. *Popular Edition.* Vol.
I. Cr. 8vo. 3s. 6d.

Shelley (Percy Bysshe). POEMS. With
an Introduction by A. CLUTTON-BROCK and
Notes by C. D. LOCOCK. *Two Volumes.*
Demy 8vo. £1 1s. net.

Smith (Adam). THE WEALTH OF
NATIONS. Edited by EDWIN CANNAN.
Two Volumes. *Second Edition. Demy*
8vo. £1 5s. net.

Stevenson (R. L.). THE LETTERS OF
ROBERT LOUIS STEVENSON. Edited
by Sir SIDNEY COLVIN. *A New Re-
arranged Edition in four volumes. Fourth
Edition. Fcap. 8vo.* Each 6s. net.

Surtees (R. S.). HANDLEY CROSS.
Illustrated. *Ninth Edition. Fcap. 8vo.*
7s. 6d. net.
MR. SPONGE'S SPORTING TOUR.
Illustrated. *Fifth Edition. Fcap. 8vo.*
7s. 6d. net.
ASK MAMMA : OR, THE RICHEST
COMMONER IN ENGLAND. Illus-
trated. *Second Edition. Fcap. 8vo.* 7s. 6d.
net.
JORROCKS'S JAUNTS AND JOLLI-
TIES. Illustrated. *Seventh Edition.*
Fcap. 8vo. 6s. net.
MR. FACEY ROMFORD'S HOUNDS.
Illustrated. *Third Edition. Fcap. 8vo.*
7s. 6d. net.
HAWBUCK GRANGE ; OR, THE SPORT-
ING ADVENTURES OF THOMAS
SCOTT, ESQ. Illustrated. *Fcap. 8vo.*
6s. net.
PLAIN OR RINGLETS? Illustrated.
Fcap. 8vo. 7s. 6d. net.
HILLINGDON HALL. With 12 Coloured
Plates by WILDRAKE, HEATH, and JELLI-
COE. *Fcap. 8vo.* 7s. 6d. net.

Tileston (Mary W.). DAILY STRENGTH FOR DAILY NEEDS. *Twenty-sixth Edition. Medium 16mo.* 3s. 6d. net.

Underhill (Evelyn). MYSTICISM. A Study in the Nature and Development of Man's Spiritual Consciousness. *Eighth Edition. Demy 8vo.* 15s. net.

Vardon (Harry). HOW TO PLAY GOLF. Illustrated. *Thirteenth Edition. Cr. 8vo.* 5s. net.

Waterhouse (Elizabeth). A LITTLE BOOK OF LIFE AND DEATH. *Twentieth Edition. Small Pott 8vo. Cloth,* 2s. 6d. net.

Wells (J.). A SHORT HISTORY OF ROME. *Seventeenth Edition.* With 3 Maps. *Cr. 8vo.* 6s.

Wilde (Oscar). THE WORKS OF OSCAR WILDE. *Fcap. 8vo. Each* 6s. 6d. net.
I. LORD ARTHUR SAVILE'S CRIME AND THE PORTRAIT OF MR. W. H. II. THE DUCHESS OF PADUA. III. POEMS. IV. LADY WINDERMERE'S FAN. V. A WOMAN OF NO IMPORTANCE. VI. AN IDEAL HUS-

BAND. VII. THE IMPORTANCE OF BEING EARNEST. VIII. A HOUSE OF POMEGRANATES. IX. INTENTIONS. X. DE PROFUNDIS AND PRISON LETTERS. XI. ESSAYS. XII. SALOMÉ, A FLORENTINE TRAGEDY, and LA SAINTE COURTISANE. XIII. A CRITIC IN PALL MALL. XIV. SELECTED PROSE OF OSCAR WILDE. XV. ART AND DECORATION.

A HOUSE OF POMEGRANATES. Illustrated. *Cr. 4to.* 21s. net.

Wood (Lieut. W. B.) and **Edmonds (Col. J. E.).** A HISTORY OF THE CIVIL WAR IN THE UNITED STATES (1861-65). With an Introduction by SPENSER WILKINSON. With 24 Maps and Plans. *Third Edition. Demy 8vo.* 15s. net.

Wordsworth (W.). POEMS. With an Introduction and Notes by NOWELL C. SMITH. *Three Volumes. Demy 8vo.* 18s. net.

Yeats (W. B.). A BOOK OF IRISH VERSE. *Fourth Edition. Cr. 8vo.* 7s. net.

PART II.—A SELECTION OF SERIES

Ancient Cities

General Editor, SIR B. C. A. WINDLE

Cr. 8vo. 6s. *net each volume*

With Illustrations by E. H. NEW, and other Artists

BRISTOL. CANTERBURY. CHESTER. DUBLIN. | EDINBURGH. LINCOLN. SHREWSBURY. WELLS and GLASTONBURY.

The Antiquary's Books

General Editor, J. CHARLES COX

Demy 8vo. 10s. 6d. *net each volume*

With Numerous Illustrations

ANCIENT PAINTED GLASS IN ENGLAND. ARCHÆOLOGY AND FALSE ANTIQUITIES. THE BELLS OF ENGLAND. THE BRASSES OF ENGLAND. THE CASTLES AND WALLED TOWNS OF ENGLAND. CELTIC ART IN PAGAN AND CHRISTIAN TIMES. CHURCHWARDENS' ACCOUNTS. THE DOMESDAY INQUEST. ENGLISH CHURCH FURNITURE. ENGLISH COSTUME. ENGLISH MONASTIC LIFE. ENGLISH SEALS. FOLK-LORE AS AN HISTORICAL SCIENCE. THE GILDS AND COMPANIES OF LONDON. THE HERMITS AND ANCHORITES OF ENGLAND. THE MANOR AND MANORIAL RECORDS. THE MEDIÆVAL HOSPITALS OF ENGLAND. OLD ENGLISH INSTRUMENTS OF MUSIC. OLD ENGLISH LIBRARIES. OLD SERVICE BOOKS OF THE ENGLISH CHURCH. PARISH LIFE IN MEDIÆVAL ENGLAND. THE PARISH REGISTERS OF ENGLAND. REMAINS OF THE PREHISTORIC AGE IN ENGLAND. THE ROMAN ERA IN BRITAIN. ROMANO-BRITISH BUILDINGS AND EARTHWORKS. THE ROYAL FORESTS OF ENGLAND. THE SCHOOLS OF MEDIEVAL ENGLAND. SHRINES OF BRITISH SAINTS.

The Arden Shakespeare

General Editor, R. H. CASE

Demy 8vo. 6s. net each volume

An edition of Shakespeare in Single Plays ; each edited with a full Introduction, Textual Notes, and a Commentary at the foot of the page.

Classics of Art

Edited by Dr. J. H. W. LAING

With numerous Illustrations. Wide Royal 8vo

THE ART OF THE GREEKS, 15s. net. THE ART OF THE ROMANS, 16s. net. CHARDIN, 15s. net. DONATELLO, 16s. net. GEORGE ROMNEY, 15s. net. GHIRLANDAIO, 15s. net. LAWRENCE, 25s. net. MICHELANGELO, 15s. net. RAPHAEL, 15s. net. REMBRANDT'S ETCHINGS, Two Vols., 25s. net. TINTORETTO, 16s. net. TITIAN, 16s. net. TURNER'S SKETCHES AND DRAWINGS, 15s. net. VELAZQUEZ, 15s. net.

The 'Complete' Series

Fully Illustrated. Demy 8vo

THE COMPLETE AMATEUR BOXER, 10s. 6d. net. THE COMPLETE ASSOCIATION FOOTBALLER, 10s. 6d. net. THE COMPLETE ATHLETIC TRAINER, 10s. 6d. net. THE COMPLETE BILLIARD PLAYER, 12s. 6d. net. THE COMPLETE COOK, 10s. 6d. net. THE COMPLETE CRICKETER, 10s. 6d. net. THE COMPLETE FOXHUNTER, 16s. net. THE COMPLETE GOLFER, 12s. 6d. net. THE COMPLETE HOCKEY-PLAYER, 10s. 6d. net. THE COMPLETE HORSEMAN, 12s. 6d. net. THE COMPLETE JUJITSUAN, 5s. net. THE COMPLETE LAWN TENNIS PLAYER, 12s. 6d. net. THE COMPLETE MOUNTAINEER, 16s. net. THE COMPLETE OARSMAN, 15s. net. THE COMPLETE PHOTOGRAPHER, 15s. net. THE COMPLETE RUGBY FOOTBALLER, ON THE NEW ZEALAND SYSTEM, 12s. 6d. net. THE COMPLETE SHOT, 16s. net. THE COMPLETE SWIMMER, 10s. 6d. net. THE COMPLETE YACHTSMAN, 16s. net.

The Connoisseur's Library

With numerous Illustrations. Wide Royal 8vo. 25s. net each volume

ENGLISH COLOURED BOOKS. ENGLISH FURNITURE. ETCHINGS. EUROPEAN ENAMELS. FINE BOOKS. GLASS. GOLDSMITHS' AND SILVERSMITHS' WORK. ILLUMINATED MANUSCRIPTS. IVORIES. JEWELLERY. MEZZOTINTS. MINIATURES. PORCELAIN. SEALS. WOOD SCULPTURE.

Handbooks of Theology

Demy 8vo

THE DOCTRINE OF THE INCARNATION, 15s. net. A HISTORY OF EARLY CHRISTIAN DOCTRINE, 16s. net. INTRODUCTION TO THE HISTORY OF RELIGION, 12s. 6d. net. AN INTRODUCTION TO THE HISTORY OF THE CREEDS, 12s. 6d. net. THE PHILOSOPHY OF RELIGION IN ENGLAND AND AMERICA, 12s. 6d. net. THE XXXIX ARTICLES OF THE CHURCH OF ENGLAND, 15s. net.

Health Series

Fcap. 8vo. 2s. 6d. net

THE BABY. THE CARE OF THE BODY. THE CARE OF THE TEETH. THE EYES OF OUR CHILDREN. HEALTH FOR THE MIDDLE-AGED. THE HEALTH OF A WOMAN. THE HEALTH OF THE SKIN. HOW TO LIVE LONG. THE PREVENTION OF THE COMMON COLD. STAYING THE PLAGUE. THROAT AND EAR TROUBLES. TUBERCULOSIS. THE HEALTH OF THE CHILD, 2s. net.

Leaders of Religion

Edited by H. C. BEECHING. *With Portraits*

Crown 8vo. 3s. *net each volume*

The Library of Devotion

Handy Editions of the great Devotional Books, well edited.
With Introductions and (where necessary) Notes

Small Pott 8vo, cloth, 3s. *net and* 3s. 6d. *net*

Little Books on Art

With many Illustrations. Demy 16mo. 5s. *net each volume*

Each volume consists of about 200 pages, and contains from 30 to 40 Illustrations,
including a Frontispiece in Photogravure

ALBRECHT DÜRER. THE ARTS OF JAPAN. BOOKPLATES. BOTTICELLI. BURNE-JONES. CELLINI. CHRISTIAN SYMBOLISM. CHRIST IN ART. CLAUDE. CONSTABLE. COROT. EARLY ENGLISH WATER-COLOUR. ENAMELS. FREDERIC LEIGHTON. GEORGE ROMNEY. GREEK ART. GREUZE AND BOUCHER. HOLBEIN. ILLUMINATED MANUSCRIPTS. JEWELLERY. JOHN HOPPNER. Sir JOSHUA REYNOLDS. MILLET. MINIATURES. OUR LADY IN ART. RAPHAEL. RODIN. TURNER. VANDYCK. VELAZQUEZ. WATTS.

The Little Guides

With many Illustrations by E. H. NEW and other artists, and from photographs

Small Pott 8vo. 4s. *net and* 6s. *net*

Guides to the English and Welsh Counties, and some well-known districts

The main features of these Guides are (1) a handy and charming form ; (2) illustrations from photographs and by well-known artists; (3) good plans and maps ; (4) an adequate but compact presentation of everything that is interesting in the natural features, history, archæology, and architecture of the town or district treated.

The Little Quarto Shakespeare

Edited by W. J. CRAIG. With Introductions and Notes

Pott 16mo. 40 *Volumes. Leather, price* 1s. 9d. *net each volume*
Cloth, 1s. 6d.

Nine Plays

Fcap. 8vo. 3s. 6d. *net*

ACROSS THE BORDER. Beulah Marie Dix. *Cr. 8vo.*

HONEYMOON, THE. A Comedy in Three Acts. Arnold Bennett. *Third Edition.*

GREAT ADVENTURE, THE. A Play of Fancy in Four Acts. Arnold Bennett. *Fifth Edition.*

MILESTONES. Arnold Bennett and Edward Knoblock. *Ninth Edition.*

IDEAL HUSBAND, AN. Oscar Wilde. *Acting Edition.*

KISMET. Edward Knoblock. *Fourth Edition.*

TYPHOON. A Play in Four Acts. Melchior Lengyel. English Version by Laurence Irving. *Second Edition.*

WARE CASE, THE. George Pleydell.

GENERAL POST. J. E. Harold Terry. *Second Edition.*

Sports Series

Illustrated.   Fcap. 8vo.   2s. net and 3s. net

ALL ABOUT FLYING, 3s. *net.* GOLF DO'S AND DONT'S. THE GOLFING SWING. HOW TO SWIM. LAWN TENNIS, 3s. *net.* SKAT- | ING, 3s. *net.* CROSS-COUNTRY SKI-ING, 5s. *net.* WRESTLING, 2s. *net.* QUICK CUTS TO GOOD GOLF, 2s. 6d. *net.*

The Westminster Commentaries

General Editor, WALTER LOCK

Demy 8vo

THE ACTS OF THE APOSTLES, 16s. *net.* AMOS, 8s. 6d. *net.* I. CORINTHIANS, 8s. 6d. *net.* EXODUS, 15s. *net.* EZEKIEL, 12s. 6d. *net.* GENESIS, 16s. *net.* HEBREWS, 8s. 6d. *net.* ISAIAH, 16s. *net.* JEREMIAH, | 16s. *net.* JOB, 8s. 6d. *net.* THE PASTORAL EPISTLES, 8s. 6d. *net.* THE PHILIPPIANS, 8s. 6d. *net.* ST. JAMES, 8s. 6d. *net.* ST. MATTHEW, 15s. *net.*

Methuen's Two-Shilling Library

Cheap Editions of many Popular Books

Fcap. 8vo

PART III.—A SELECTION OF WORKS OF FICTION

Bennett (Arnold)—

CLAYHANGER, 8s. *net.* HILDA LESSWAYS, 8s. 6d. *net.* THESE TWAIN. THE CARD. THE REGENT: A Five Towns Story of Adventure in London. THE PRICE OF LOVE. BURIED ALIVE. A MAN FROM THE NORTH. THE MATADOR OF THE FIVE TOWNS. WHOM GOD HATH JOINED. A GREAT MAN: A Frolic. *All 7s. 6d. net.*

Birmingham (George A.)—

SPANISH GOLD. THE SEARCH PARTY. LALAGE'S LOVERS. THE BAD TIMES. UP, THE REBELS. *All 7s. 6d. net.*

Burroughs (Edgar Rice)—

TARZAN OF THE APES, 6s. *net.* THE RETURN OF TARZAN, 6s. *net.* THE BEASTS OF TARZAN, 6s. *net.* THE SON OF TARZAN, 6s. *net.* JUNGLE TALES OF TARZAN, 6s. *net.* TARZAN AND THE JEWELS OF OPAR, 6s. *net.* TARZAN THE UNTAMED, 7s. 6d. *net.* A PRINCESS OF MARS, 6s. *net.* THE GODS OF MARS, 6s. *net.* THE WARLORD OF MARS, 6s. *net.*

Conrad (Joseph). A SET OF SIX. *Fourth Edition. Cr. 8vo. 7s. 6d. net.*
VICTORY: AN ISLAND TALE. *Sixth Edition. Cr. 8vo. 9s. net.*

Corelli (Marie)—

A ROMANCE OF TWO WORLDS, 7s. 6d. *net.* VENDETTA: or, The Story of One Forgotten, 8s. *net.* THELMA: A Norwegian Princess, 8s. 6d. *net.* ARDATH: The Story of a Dead Self, 7s. 6d. *net.* THE SOUL OF LILITH, 7s. 6d. *net.* WORMWOOD: A Drama of Paris, 8s. *net.* BARABBAS: A Dream of the World's Tragedy, 8s. *net.* THE SORROWS OF SATAN, 7s. 6d. *net.* THE MASTER-CHRISTIAN, 8s. 6d. *net.* TEMPORAL POWER: A Study in Supremacy, 6s. *net.* GOD'S GOOD MAN: A Simple Love Story, 8s. 6d. *net.* HOLY ORDERS: The Tragedy of a Quiet Life, 8s. 6d. *net.* THE MIGHTY ATOM, 7s. 6d. *net.* BOY: A Sketch, 7s. 6d. *net.* CAMEOS, 6s. *net.* THE LIFE EVERLASTING, 8s. 6d. *net.*

Doyle (Sir A. Conan). ROUND THE RED LAMP. *Twelfth Edition. Cr. 8vo. 7s. 6d. net.*

Hichens (Robert)—

TONGUES OF CONSCIENCE, 7s. 6d. *net.* FELIX: Three Years in a Life, 7s. 6d. *net.* THE WOMAN WITH THE FAN, 7s. 6d. *net.* BYEWAYS, 7s. 6d. *net.* THE GARDEN OF ALLAH, 8s. *net.* THE CALL OF THE BLOOD, 8s. 6d. *net.* BARBARY SHEEP, 6s. *net.* THE DWELLERS ON THE THRESHOLD, 7s. 6d. *net.* THE WAY OF AMBITION, 7s. 6d. *net.* IN THE WILDERNESS, 7s. 6d. *net.*

Hope (Anthony)—
A Change of Air. A Man of Mark. The Chronicles of Count Antonio. Simon Dale. The King's Mirror. Quisanté. The Dolly Dialogues. Tales of Two People. A Servant of the Public. Mrs. Maxon Protests. A Young Man's Year. Beaumaroy Home from the Wars. *All 7s. 6d. net.*

Jacobs (W. W.)—
Many Cargoes, 5s. *net* and 2s. 6d. *net*. Sea Urchins, 5s. *net* and 3s. 6d. *net*. A Master of Craft, 5s. *net*. Light Freights, 5s. *net*. The Skipper's Wooing, 5s. *net*. At Sunwich Port, 5s. *net*. Dialstone Lane, 5s. *net*. Odd Craft, 5s. *net*. The Lady of the Barge, 5s. *net*. Salthaven, 5s. *net*. Sailors' Knots, 5s. *net*. Short Cruises, 5s. *net*.

London (Jack). WHITE FANG. *Ninth Edition. Cr. 8vo. 7s. 6d. net.*

McKenna (Stephen)—
Sonia : Between Two Worlds, 8s. *net*. Ninety-Six Hours' Leave, 7s. 6d. *net*. The Sixth Sense, 6s. *net*. Midas & Son, 8s. *net*.

Malet (Lucas)—
The History of Sir Richard Calmady : A Romance. The Wages of Sin. The Carissima. The Gateless Barrier. Deadham Hard. *All 7s. 6d. net.*

Mason (A. E. W.). CLEMENTINA. Illustrated. *Ninth Edition. Cr. 8vo. 7s. 6d. net.*

Maxwell (W. B.)—
Vivien. The Guarded Flame. Odd Lengths. Hill Rise. The Rest Cure. *All 7s. 6d. net.*

Oxenham (John)—
A Weaver of Webs. Profit and Loss. The Song of Hyacinth, and Other Stories. Lauristons. The Coil of Carne. The Quest of the Golden Rose. Mary All-Alone. Broken Shackles. "1914." *All 7s. 6d. net.*

Parker (Gilbert)—
Pierre and his People. Mrs. Falchion. The Translation of a Savage. When Valmond came to Pontiac : The Story of a Lost Napoleon. An Adventurer of the North : The Last Adventures of 'Pretty Pierre.' The Seats of the Mighty. The Battle of the Strong : A Romance of Two Kingdoms. The Pomp of the Lavilettes. Northern Lights. *All 7s. 6d. net.*

Phillpotts (Eden)—
Children of the Mist. Sons of the Morning. The River. The American Prisoner. Demeter's Daughter. The Human Boy and the War. *All 7s. 6d. net.*

Ridge (W. Pett)—
A Son of the State, 7s. 6d. *net*. The Remington Sentence, 7s. 6d. *net*. Madame Prince, 7s. 6d. *net*. Top Speed, 7s. 6d. *net*. Special Performances, 6s. *net*. The Bustling Hours, 7s. 6d. *net*.

Rohmer (Sax)—
The Devil Doctor. The Si-Fan Mysteries. Tales of Secret Egypt. The Orchard of Tears. The Golden Scorpion. *All 7s. 6d. net.*

Swinnerton (F.). SHOPS AND HOUSES. *Third Edition. Cr. 8vo. 7s. 6d. net.*
SEPTEMBER. *Third Edition. Cr. 8vo. 7s. 6d. net.*

Wells (H. G.). BEALBY. *Fourth Edition. Cr. 8vo. 7s. 6d. net.*

Williamson (C. N. and A. M.)—
The Lightning Conductor : The Strange Adventures of a Motor Car. Lady Betty across the Water. Scarlet Runner. Lord Loveland discovers America. The Guests of Hercules. It Happened in Egypt. A Soldier of the Legion. The Shop Girl. The Lightning Conductress. Secret History. The Love Pirate. *All 7s. 6d. net.* Crucifix Corner. 6s. *net*.

Methuen's Two-Shilling Novels

Cheap Editions of many of the most Popular Novels of the day

Write for Complete List

Fcap. 8vo

www.ingramcontent.com/pod-product-compliance
Lightning Source LLC
Chambersburg PA
CBHW020526270326
41927CB00006B/466